# First Workshop on Language Grounding for Robotics 2017

Held at ACL 2017

Vancouver, Canada
3 August 2017

ISBN: 978-1-5108-4578-7

ACL 2017

# The First Workshop on Language Grounding for Robotics

## Proceedings of the Workshop

August 3, 2017
Vancouver, Canada

# Introduction

After the remarkable successes of recent work visually grounded models of language, the embodied and task-oriented aspects of language learning stand as a natural next challenge. As autonomous robotic agents become increasingly capable and are deployed to progressively more complex environments, expressive, accessible interfaces are becoming essential to realizing the potential of such technologies. Natural language is immediately available to non-expert users and expressive enough to represent complex actions and plans. Can we give instructions to robotic agents to assist with navigation and manipulation tasks in remote settings? Can we talk to robots about the surrounding visual world, and help them interactively learn the language needed to finish a task? To build robots that we can converse with in our homes, offices, hospitals, and warehouses, it is essential that we develop new techniques for linking language to action in the real world.

While the opportunity is clear, enabling effective interaction between users and autonomous agents requires addressing some of the core open challenges in NLP while studying new domains and tasks. This workshop aims to explore these challenges, bringing together members of the NLP, robotics, and vision communities to focus on language grounding in robots and other interactive goal-driven systems. The program features twelve new articles and seven cross-submissions from related areas, to be presented as both posters and talks. We are also excited to host remarkable invited speakers, including Regina Barzilay, Joyce Chai, Karl Moritz Hermann, Hadas Kress-Gazit, Terence Langendoen, Percy Liang, Ray Mooney, Nicholas Roy, Stefanie Tellex and Jason Weston.

We thank the program committee, the ACL workshop chairs Wei Xu and Jonathan Berant, the invited speakers, and our sponsors DeepMind and Facebook.

—Mohit Bansal, Cynthia Matuszek, Jacob Andreas, Yoav Artzi and Yonatan Bisk, organizers

**Organizers:**

Mohit Bansal, UNC Chapel Hill
Cynthia Matuszek, UMBC
Jacob Andreas, UC Berkeley
Yoav Artzi, Cornell
Yonatan Bisk, ISI, USC

**Program Committee:**

Antoine Bordes, Facebook AI Research
Ankur Parikh, Google Research
Devi Parikh, Georgia Tech
Dhruv Batra, Georgia Tech
Dieter Fox, Univ of Washington
Dipendra Misra, Cornell
Edward Grefenstette, Google DeepMind
Eunsol Choi, Univ of Washington
Hadas Kress-Gazit, Cornell
Hannaneh Hajishirzi, Univ of Washington
Hongyuan Mei, JHU
Jason Weston, Facebook AI Research
Jayant Krishnamurthy, AI2
Jesse Dodge, CMU
Jesse Thomason, UT Austin
Jivko Sinapov, UT Austin
Jonathan Berant, Tel-Aviv
Joyce Chai, MSU
Julia Hockenmaier, UIUC
Karthik Narasimhan, MIT
Kenton Lee, Univ of Washington
Licheng Yu, UNC Chapel Hill
Lisa A Hendricks, UC Berkeley
Lucy Vanderwende, MSR
Marcus Rohrbach, UC Berkeley
Mark Yatskar, Univ of Washington
Matthew Walter, TTI-Chicago
Raia Hadsell, Google DeepMind
Ray Mooney, UT Austin
Siva Reddy, Univ of Edinburgh
Subhashini Venugopalan, UT Austin
Thomas Kollar, Amazon
Tom Kwiatkowski, Google Research
Tom Williams, Tufts
Yejin Choi, Univ of Washington

# Table of Contents

# Workshop Program

# Grounding Language for Interactive Task Learning

Peter Lindes, Aaron Mininger, James R. Kirk, and John E. Laird
Computer Science and Engineering
University of Michigan, Ann Arbor, MI 48109-2121
{plindes, mininger, jrkirk, laird}@umich.edu

## Abstract

This paper describes how language is grounded by a comprehension system called Lucia within a robotic agent called Rosie that can manipulate objects and navigate indoors. The whole system is built within the Soar cognitive architecture and uses Embodied Construction Grammar (ECG) as a formalism for describing linguistic knowledge. Grounding is performed using knowledge from the grammar itself, from the linguistic context, from the agent's perception, and from an ontology of long-term knowledge about object categories and properties and actions the agent can perform. The paper also describes a benchmark corpus of 200 sentences in this domain, along with test versions of the world model and ontology, and gold-standard meanings for each of the sentences. The benchmark is contained in the supplemental materials.

## 1 Introduction

This paper considers language grounding within the context of Interactive Task Learning (ITL; Laird et al., 2017), where the goal is to teach an intelligent agent new tasks and extend existing tasks through natural language instruction by a human teacher. This kind of instruction has been done with an agent called Rosie (Mohan and Laird, 2014). Rosie has been interfaced to a tabletop robot arm and a mobile robot that can navigate and perform tasks in an indoor environment. We discuss the techniques and strategies used to ground the natural language input to the knowledge the agent has about the world and its own capabilities.

Rosie can perform simple manipulation tasks like *'Pick up the green sphere.'* or *'Put that in the pantry.'*, simple navigation tasks like *'Go to the kitchen.'* or *'Follow the right wall.'* and more complex tasks like *'Fetch a soda.'* or *'Deliver the package to the main office.'* It can also understand descriptions of objects and answer simple questions about its world.

In the work described here, the language comprehension is performed by a system called Lucia. Lucia runs as a part of Rosie, and is under continuing development. Previous work (Lindes and Laird, 2016; 2017a; 2017b) described some aspects of Lucia, but here we describe in some detail how Lucia does language grounding within Rosie. We also provide a benchmark that may be useful for comparing language grounding systems for robots.

### 1.1 Research Context

Our research is embedded in a cognitive modelling approach (Laird et al., 2012). This affects our goals and methods in three ways. First, we attempt to implement all aspects of Rosie's intelligence, including language comprehension, task planning, dialog interaction, etc., within a single agent built on an architecture designed to model general principles of human cognition. Specifically, we use the Soar cognitive architecture (Laird, 2012).

Second, we wish to apply a theory of linguistic or grammatical knowledge, based on cognitive linguistics research, that combines syntactic and semantic knowledge in a single integrated grammar. Lucia uses Embodied Construction Grammar (ECG; Feldman et al., 2009; Bergen and Chang, 2013) as that theory. This theory has the potential to scale to cover much variation in human language use, as well as relating to the complexities of human conceptual models of the world.

Third, with Lucia we seek to build a language comprehension process that conforms to psycholinguistic research on incremental human processing (Christiansen and Chater, 2016) and draws directly on all the contextual knowledge the agent has. This approach may have an advantage in meeting human expectations of how natural language will be understood by the agent.

This cognitive modeling approach differs from many other approaches to language comprehension and language grounding for robots found in the literature. For example, a number of research-

1

*Proceedings of the First Workshop on Language Grounding for Robotics*, pages 1–9,
Vancouver, Canada, July 30 - August 4, 2017. ©2017 Association for Computational Linguistics

ers have built systems to satisfy the need for grounded language in a robot, including Steels and Hild (2012), Tellex et al. (2014), and Eppe et al. (2016). However, none of those systems conforms to the three aspects of a cognitive modeling approach outlined above. Our work is beginning to explore whether the cognitive approach can provide more robust language understanding.

Lucia could be characterized as a semantic parser using textual and visual context since it takes natural language utterances in, uses knowledge of the surrounding text and the objects seen by the agent, and produces semantic representations as its output. However, it differs in many respects from many other "semantic parsers."

For example, Zettlemoyer and colleagues (Zettlemoyer and Collins, 2007; Artzi and Zettlemoyer, 2013) have built systems for learning mappings from utterances to logical forms. Berant et al. (2013) learn a system to map questions to answers in relation to a large database. Wang et al. (2016) report on a system for learning the language needed to instruct a computer to perform certain tasks.

All these systems have produced impressive results. Our results with Lucia are complementary in two important ways. The meaning representations they produce are not just logical forms but are connected to the perception and action knowledge of a fully embodied agent. Lucia also satisfies the cognitive modelling constraints we outlined above, which none of these other systems attempt to do. The richness of this variety of different approaches should help advance future research. Our contribution is to show something of what is possible using a cognitive model embodied in a robotic agent.

### 1.2 Theoretical Background

Rosie is built within the Soar cognitive architecture. The Soar architecture has a working memory with information about its current perception of the world as well as its internal goals and state. A procedural memory contains production rules that represent knowledge of how to perform internal and external actions. A long-term semantic memory holds knowledge of the categories of objects the agent knows about, what properties these objects can have, and the actions the agent knows how to perform. Dynamic operation in Soar consists of a series of decision cycles, where in each cycle a single operator is selected and applied, and that operator influences which production rules fire to make changes in working memory or initiate external actions during that cycle.

In order to learn new tasks from instruction, Rosie must have a natural language understanding capability. That capability must be able to produce a meaning structure for each input utterance that is *grounded* to the agent's perception, action capabilities, and general world knowledge. By grounded we mean that the resulting meaning structure refers directly to the agent's internal representation of objects, of the actions it knows how to perform, and of any other relevant knowledge the agent has, such as spatial relations between objects.

Several approaches to language comprehension in Soar have been used previously (Lehman et al., 1991; Mohan et al., 2012; Mohan and Laird, 2014; Kirk et al. 2016). More recently a language comprehension system in Soar called Lucia (Lindes and Laird, 2016; 2017a) has been developed. In addition to the general cognitive abilities inherent in Soar, Lucia uses a cognitive theory of language called Embodied Construction Grammar (ECG; Feldman et al., 2009; Bergen and Chang, 2013).

The ECG grammar formalism (Bryant, 2008) defines a grammar in terms of two kinds of items: *constructions* and *schemas*. A construction is a pairing of form and meaning. Some constructions match individual lexical items and others match one or more constituents in a recursive hierarchy. Each construction describes its meaning in terms of schemas.

A schema can be thought of as a feature bundle. It defines a data structure that has a type name and one or more *roles*, or slots, to hold information. A construction defines what schema is to be evoked when it is recognized and how to fill the roles of that schema. Roles can be filled with constants provided in a construction or from the meaning structures of the constituents of a construction, gradually building a complex hierarchical meaning structure as each sentence is comprehended.

As an example of Lucia's comprehension, consider the sentence *'Pick up the green sphere.'* Figure 1 shows the data structures Lucia builds to comprehend this sentence. The blue rectangles represent the constructions that were recognized, the green ovals are the meaning schemas that were

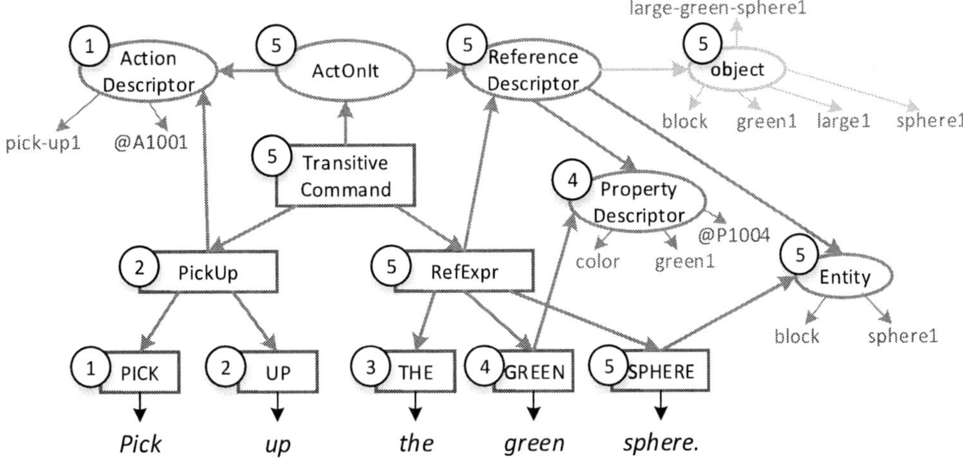


Figure 1: Comprehension of a simple sentence. (Adapted from Lindes and Laird, 2016.)

```
construction TransitiveCommand          schema Action
    subcase of Imperative                   roles
    constructional
        constituents                            action
            verb: ActionVerb                    location
            object: RefExpr
    meaning: ActOnIt                    schema ActOnIt
        constraints                         subcase of Action
            self.m.action <--> verb.m       roles
            self.m.object <--> object.m          object
```

Figure 2: ECG example. (Adapted from Lindes and Laird, 2016.)

instantiated, and the orange and red items represent information derived from grounding.

In the example in Figure 1, there are five word cycles, one for each word in the input sentence. The circled numbers in the figure indicate during which word cycle each construction or schema was instantiated or each grounding link formed.

Figure 2 shows the ECG form of the construction for TransitiveCommand, which produces the blue rectangle with the same name in the figure, along with the schemas used to build its meaning. The construction specifies the types of constituents needed to trigger the instantiation of this construction, the type of the schema to be evoked, and constraints for mapping the meanings of its constituents to the slots in the meaning structure.

Lucia stores all its linguistic knowledge in Soar's procedural memory, thus avoiding the overhead of retrieving this knowledge from se-mantic memory as the earlier systems generally do.

## 1.3 Overview

This paper concentrates on the methods used for grounding language in the Lucia system. Supplemental materials describe a benchmark to enable evaluating other systems against the same sentence corpus we have used for testing and a set of gold-standard meanings for those sentences. We have found a lack of published material to compare systems for language grounding in robots, and our intention is that this benchmark can be one attempt to fill this gap.

The following sections discuss our approach to grounding, give some examples of language used for ITL, and describe the grounding processes in Lucia. Finally we describe the files contained in the benchmark, which are submitted as supplemental material with this paper.

## 2    Grounded Meanings

Human interaction with robots using natural language often needs language to be grounded to the agent's perceptions of the physical world and its knowledge of its own action capabilities.

Several projects have used the Rosie system to explore ITL. Mohan et al. (2012) discuss interactive methods for learning words that are grounded in the agent's physical environment and actions with a table-top robotic arm. Kirk and Laird (2016) report using interactive instruction to teach Rosie to understand and play new games. The application to games raises many issues with grounding language in hypothetical settings, but this paper does not consider this aspect. Mininger and Laird (2016) extend the table-top version of Rosie to one that can navigate in an indoor environment and comprehend language about objects that are unseen or unknown. These projects have contributed much to task learning, but their language comprehension systems are ad-hoc.

Lucia (Lindes and Laird, 2016; 2017a; 2017b) is a comprehension system built within the same Rosie agent and using the Soar architecture, but it also is built around the ECG theory of language. Its linguistic knowledge is written by hand in the ECG formalism (Bryant, 2008) and translated automatically into Soar production rules. Since all this knowledge is procedural and does not have to be retrieved dynamically from long-term memory, Lucia simulates skilled comprehension in simulated time close to human real-time performance. It uses a human-like incremental processing system, distinct from the best-fit over a whole sentence approach used by other ECG systems (Bryant, 2008). In what follows we look at how well Lucia succeeds in grounding language within the constraints imposed by ECG and incremental processing.

What does it mean to ground natural language in this context? The comprehension of each input sentence must produce a meaning structure in working memory that is sufficient for the agent to use its knowledge of perception and action to perform the internal and external actions the instructor intended. In ITL the interaction process may include requests from the agent for additional information or clarification.

In this paper we do not consider the details of how the agent's perception and action work. Rather we assume that before trying to comprehend a given input utterance, the agent already has knowledge about what its vision system currently perceives in the world. It also has knowledge in long-term memory about what actions it can perform. Knowledge of actions can be either built into the agent or learned through interaction. In either case, the perception and action concepts which the agent grounds to physical percepts or motor control programs are represented by internal symbols shared by the linguistic and robotic parts of the agent. Thus we are concerned here with grounding the natural language to these internal symbols and compositions of them.

In order to ground the meaning of a sentence, each linguistic unit involved must be grounded, including words, phrases, clauses, and complete sentences. To comprehend *'Pick up the green sphere.'* as shown in Figure 1, *pick up* must be grounded to an action the agent knows how to perform, *the green sphere* must be grounded to a specific object that the agent sees in its current environment, and these two meanings must be composed into a sentence-level meaning that can produce an actionable "message" which tells Rosie what action to take. Along the way the meanings of individual words like *green* and *sphere* must be grounded to the corresponding properties in the agent's long-term knowledge that are required to find the object in its perceived scene.

## 3    Language Used for Interactive Task Learning

In ITL the agent starts with sufficient linguistic and operational knowledge to perform some tasks, but then needs to learn new tasks and extensions to known tasks through interaction with a human instructor (Laird et al., 2012; Mohan et al., 2012; Mininger and Laird, 2016). In this section we give some examples of the language input involved in learning a few example tasks. Although the interaction also involves requests from Rosie to the instructor, we consider only language comprehension and not production here.

Assume that at first the agent knows the primitive manipulation command to *pick up* an object in its visual field and another to *put* or *put down* that object in one of its known locations. Now we can instruct it to learn the verb *move* with an interaction that includes the following sequence of instructions, interspersed with agent responses that are not shown.

(1) Sentences for teaching the verb *move*

a.     `Move the red block to the left of the orange block.`

b.     `The goal is that the red block is to the left of the orange block.`

c.     `Pick up the red block.`

d.     `Put the red block to the left of the orange block.`

e.     `The task is done.`

In (1a) a command to perform an unknown task is given, and the agent asks for help. Then the instructor states the end goal of the task (1b). In some cases this may be sufficient, and the agent may be able to perform the reasoning needed to plan a sequence of actions to perform the task. In (1) we show a case where the agent asks for more help after (1b), and the instructor gives a sequence of known commands needed to complete the task (1c-e). Rosie can then remember the goal, its relation to the original *move* command, and the sequence of steps so that if given another *move* command in the future it can perform it unaided. Other examples of similar interactions are given in (2) through (4).

(2) Sentences for teaching the *discard* task

a.     `Discard the green box.`

b.     `The goal is that the green box is in the trash.`

c.     `Pick up the green box.`

d.     `Put the green box in the trash.`

e.     `The task is finished.`

(3) Sentences for teaching a *deliver* task

a.     `Deliver the box to the main office.`

b.     `The goal is that the box is in the office.`

c.     `Pick up the box.`

d.     `Go to the main office.`

e.     `Put down the box.`

f.     `You are done.`

(4) Sentences for teaching a *fetch* task

a.     `Fetch a stapler.`

b.     `The goal is that the stapler is in the starting location.`

c.     `Remember the current location as the starting location.`

d.     `Find the stapler.`

e.     `Pick up the stapler.`

f.     `Go to the starting location.`

g.     `Put down the stapler.`

h.     `The task is over.`

These examples illustrate the kinds of sequences involved in ITL, but do not represent the full linguistic range of the system. The benchmark described below contains a corpus of 200 sentences that apply to the object manipulation and indoor navigation domains, as well as to learning complex tasks in these domains. These sentences have been designed by hand to accomplish three purposes: provide the information needed to achieve our ITL goals, say things in a way that seems natural to humans, and experiment with different linguistic forms.

The entire corpus includes declarative sentences to describe objects or relations, commands for object manipulation and indoor navigation, conditional *if/then* commands, commands with *until* clauses, sentences about goals and task progress, and questions to Rosie about its knowledge of the world. Unrestricted human interaction with the agent might well produce many additional linguistic forms we have not yet considered. As a group, the 200 sentences provide a number of comprehension challenges, including lexical, syntactic, and semantic ambiguities (Lindes and Laird, 2017b).

## 4    The Grounding Process in Lucia

This section examines how Lucia grounds words, phrases, clauses, and sentences, eventually producing a *message* to the operational part of Rosie for each sentence it comprehends. We describe the knowledge sources used to provide information for the grounding, give an overview of the comprehension process in Lucia, and describe the various grounding processes.

### 4.1    Knowledge Sources

Information for grounding the various linguistic units comes from four sources: the ECG grammar, the current state of the comprehension, the current perceived visual scene, and an ontology of classes, properties, and actions.

**The ECG grammar:** Lucia uses a grammar built by hand in the ECG language. An off-line program translates the constructions and schemas in the grammar into Soar production rules (Lindes and Laird, 2016). As the comprehension proceeds, these rules fire at appropriate times to instantiate constructions and schemas and fill the schema roles whose fillers are defined in the grammar. Additional Soar rules not built from the grammar fill in grounding information from the other knowledge sources.

**The comprehension state**: The Lucia comprehension system works incrementally, doing as much processing as possible as each individual word comes in (Lindes and Laird, 2017a). In doing so it recognizes lexical and phrasal constructions and builds a hierarchy of their instances. Schemas are also instantiated and filled as soon as possible and attached to the constructions. At any point in time the comprehender has a stack of construction instances built so far that have not yet been incorporated as constituents in higher level constructions, and each of these has its attached meaning. Thus the rules that are trying to ground any new meaning being constructed can draw on the knowledge contained in this comprehension state as one of their information sources.

**The world model**: Rosie has a scene graph is brought into Soar's working memory by its visual perception system and thus is available to Lucia and to the part of Rosie that implements actions. Each object is identified by a unique identifier and has category, color, size, and shape properties set by the visual perception system. In addition to objects, the world model contains information about spatial relations between objects, an indicator of which object the instructor is currently pointing to, and a special object to represent the robot itself. This world model is a key source of knowledge for grounding language.

**The ontology:** The objects and relations in the world model are in working memory and can change as Rosie proceeds through a task. We also need fixed knowledge to represent categories, property values, and actions. The source for this kind of knowledge is an ontology stored in Soar's long-term semantic memory. This knowledge, like the world model, is shared between Lucia and the operational rules in Rosie.

## 4.2 Lucia Comprehension Overview

Lucia processes a sentence word-by-word and left-to-right. A number of Soar operators are selected and applied during each word cycle. By the end of the word cycle, the comprehension state will have the lexical construction for that word, larger phrasal constructions that combine it as appropriate with items previously on the stack, and

grounded meaning schemas corresponding to these new constructions.

Each word known to the system has one or more lexical constructions in the grammar. If a word has multiple senses, each of these constructions is instantiated at first, and later processes select the correct one for the current context (Lindes and Laird, 2017b). Phrasal constructions combine constituents into higher level constructions. As each construction is instantiated, it evokes a schema to represent its meaning. Along the way, the various forms of grounding are performed.

## 4.3 Grounding Referring Expressions

We define a *referring expression* as a linguistic unit meant to describe some object that the system can know about. The general construction for a referring expression is called a `RefExpr`, and its meaning is represented by a `RefDesc`, or *reference descriptor*. These are built up as words are being processed. A `RefExpr` can consist of a simple pronoun, like *it* or *this*, a noun phrase like *the green sphere*, or a more complex expression like *the green rectangle to the left of the large green rectangle* or *a green block that is on the stove*. As the individual words in the expression are processed, the complete `RefExpr` and its `RefDesc` meaning are gradually built up.

A common noun generates a schema that represents some class of object, and sets the roles of this schema to identifiers for the category and/or shape of that class of objects. An adjective generates a schema to describe a property class and a value for that property, such as the color *green*. A determiner sets whether the expression is *definite* or *indefinite*. At this lexical level, part of the grounding is performed by instantiating the grammatical knowledge incorporated in the lexical constructions. Then an operator is selected to retrieve information about categories and properties from the ontology.

As soon as a complete noun phrase has been built, an operator takes the `RefDesc` that has been assembled and searches in the world model for one or more objects that match the description. Similarly, a pronoun is grounded by an operator that deals with pronouns. The object that is found from this grounding is then set as the referent of the `RefDesc`.

More complex referring expressions are grounded in several steps. For instance, *the green rectangle to the left of the large green rectangle* causes several phrase-level grounding operations in addition to the lexical ones. The phrase *the green rectangle* will be grounded to the set of two green rectangles in our test world model, while the *large green rectangle* will be grounded to a single object. The preposition *to the left of* is looked up in the ontology to see what kind of relation it represents. Finally that relation is used to select from the possible green rectangles the one which satisfies the complete expression.

A complex referring expression like *a green block that is on the stove* also requires several grounding steps. First when *a green block* is grounded, a set of several objects that match is formed. Next *that* is recognized as a relative pronoun, and connected to the preceding noun phrase. The phrase *on the stove* finds the stove and a relation to it, which is then applied to the set of objects found for the noun phrase that *that* is attached to, resulting in a single object that satisfies the whole expression.

### 4.4 Grounding and Attaching Prepositional Phrases

Consider the full sentence: *'Move the green rectangle to the left of the large green rectangle to the pantry.'* Above we looked at grounding *the green rectangle to the left of the large green rectangle* as an isolated expression. Within the complete sentence, however, things get more complicated. The prepositional phrase with *to the left of* could attach to the previous noun phrase, but it could also attach to the verb as its target location. Then later on, as the incremental processing proceeds, we get to the phrase *to the pantry*. Where should this be attached?

Lucia has a strategy for resolving issues of this sort within its incremental, single-path parsing strategy context (Lindes and Laird, 2017a). When the first prepositional phrase has been assembled, two attachment sites are considered: the immediately preceding noun phrase or the previous verb. If the verb is one that requires a target location, such as *put* or *move*, the prepositional phrase will be attached to the verb. If the verb is one like *pick up* that does not require a target location, the phrase will be attached to the preceding noun phrase.

When the second prepositional phrase has been processed, however, we have a problem. The verb already has a target location attached, and the noun phrase before that has been hidden under the construction that makes that attachment. Here Lucia uses a strategy called *local repair*. A *snip* operation disconnects the first prepositional phrase from the verb, and then it is reattached to *the green rectangle,* and that complete expression is regrounded. Now the phrase *to the pantry* can be attached to the verb as its target location. This local repair operation is described more fully by Lindes and Laird (2017b).

### 4.5 Grounding Full Sentences

The ECG grammar provides constructions that combine the lexical items for verbs with the referring expressions that form the verb's arguments to form complete sentences. These are often called argument structure constructions. Verbs describing actions are grounded by looking up their identifiers in the ontology to connect to the actions the agent knows how to perform. This lookup provides the referring link to the agent's knowledge of how to perform the given action.

Once the comprehension process has recognized a complete sentence as a single construction, an *interpretation* process is performed. This process converts the top-level meaning structure produced by the language comprehension system into a grounded, actionable *message* for Rosie to act on. Every message has a type field, plus other arguments depending on its type. Here we give brief descriptions of these messages; more detail can be found in the supplemental materials.

In the example in Figure 1, the `ActOnIt` schema is interpreted to form a message of type `command` with arguments `pick-up1` and `large-green-sphere1`.

Declarative sentences produce `object-description` messages. This message type has an object argument to indicate the object being described and a property argument showing the property to be assigned to the object. All action commands produce `command` messages. Each `command` message has an action argument, most have an object argument, and others have varying arguments.

The command *'If you see the soda then pick it up.'* illustrates a conditional command which produces a "conditional" message an action for the *then* clause and a condition for the *if* clause. The condition must be met first, and the action will be performed when the condition is met. A command can also have an *until* clause describing a condition to terminate the action, as in *'Explore until you see a soda.'*

Lucia and Rosie understand several types of questions, as shown in (5).

(5)   Some questions that Rosie can understand

a.   `Is the large sphere green?`

b.   `Is the small orange triangle behind the green sphere?`

c.   `What is inside the pantry?`

d.   `Where is the red triangle?`

e.   `What color is the large sphere?`

f.   `What shape is this?`

Such questions produce various forms of question messages, with arguments to define the objects, properties, or relations being asked about.

## 5   Benchmark

We have assembled various data items discussed in this paper into a package to be submitted as supplementary material. We hope this package, which we are calling *The University of Michigan Robot Language Benchmark #1*, will be useful to other researchers as a benchmark against which to evaluate their systems for robot language grounding as we are using it to evaluate and continue to develop Lucia.

The supplementary materials are contained in a file called UMRLB-1_v0.1.zip containing the files listed in Table 1. The "-1" indicates that we expect there to be others in the future, and the "_v0.1" indicates the specific version. The files containing data structures are in the industry standard JSON format to make them easily machine-readable across many systems.

### Acknowledgments

The work described here was supported by the AFOSR under Grant Number FA9559-15-1-0157. The views and conclusions contained in this document are those of the authors and should not be interpreted as representing the official policies, either expressly or implied, of the AFOSR or the U.S. Government.

| File Name | Description |
|---|---|
| UMRLB-1 .pdf | A document describing in detail the files in the benchmark and their meanings. |
| Sentences.txt | The corpus of 200 sentences, grouped by their linguistic types. |
| World.json | A definition of a particular snapshot of the world perceived by a robot that can be used to ground linguistic expressions. |
| Ontology.json | An ontology defining properties of perceived objects and robot actions. |
| GoldStandard .json | A file giving the gold-standard meaning for each sentence in the corpus, along with other metadata. |

Table 1: Files included in the Benchmark

## References

Yoav Artzi and Luke Zettlemoyer. 2013. Weakly Supervised Learning of Semantic Parsers for Mapping Instructions to Actions. *Transactions of the Association for Computational Linguistics, 1, 49-62.*

Jonathan Berant, Andrew Chou, Roy Frostig, and Percy Liang. 2013. Semantic Parsing on Freebase from Question-Answer Pairs. *EMNLP 2013: Conference on Empirical Methods in Natural Language Processing.*

Benjamin Bergen and Nancy Chang. 2013. Embodied construction grammar. In Thomas Hoffman and Graeme Trousdale, editors, *The Oxford Handbook of Construction Grammar*, Oxford University Press, Oxford, chapter 10, pages 168–190.

John Edward Bryant. 2008. *Best-fit Constructional Analysis*. Ph.D. thesis, Berkeley, CA, USA.

Morten H. Christiansen and Nick Chater. 2016. The Now-or-Never bottleneck: A fundamental constraint on language. *Behavior and Brain Science.* doi:10.1017/S0140525X1500031X.

Manfred Eppe, Sean Trott, Vivek Raghuram, Jerome Feldman, and Adam Janin. 2016. Application independent and integration-friendly natural language understanding. In Christoph Benzmüller, Geoff Sutcliffe, and Raul Rojas, editors, GCAI 2016. *2nd Global Conference on Artificial Intelligence.* EasyChair, volume 41 of EPiC Series in Computing, pages 340–352.

Jerome Feldman, Ellen Dodge, and John Bryant. 2009. Embodied construction grammar. In Bernd Heine and Heiko Narrog, editors, *The Oxford Handbook of Linguistic Analysis*, Oxford University Press, Oxford https://doi.org/10.1093/oxfordhb/9780199544004.013.0006.

James Kirk, Aaron Mininger, and John Laird. 2016. Learning task goals interactively with visual demonstrations. *Biologically Inspired Cognitive Architectures* 18:1–8.

James R Kirk and John E Laird. 2016. Learning general and efficient representations of novel games through interactive instruction. *Adv. Cogn. Syst 4*.

John Laird, Keegan Kinkade, Shiwali Mohan, and Joseph Xu. 2012. Cognitive robotics using the soar cognitive architecture. In *Workshops at the Twenty-Sixth AAAI Conference on Artificial Intelligence, Cognitive Robotics*. https://www.aaai.org/ocs/index.php/WS/AAAIW12/paper/view/5221.

John E. Laird, K. Gluck, John Anderson, Ken Forbus, O. Jenkins, Christian Lebiere, D. D. Salvucci, Matthias Scheutz, A. Thomaz, G. Trafton, R. E. Wray, Shiwali Mohan, and James R. Kirk. 2017. Interactive Task Learning, *IEEE Intelligent Systems*, In press.

John E Laird. 2012. *The Soar cognitive architecture*. MIT Press.

Fain Lehman, Richard L Lewis, and Allen Newell. 1991. Integrating knowledge sources in language comprehension. In Proceedings of the Thirteenth *Annual Conference of the Cognitive Science Society*. pages 461–466.

Peter Lindes and John E Laird. 2016. Toward Integrating Cognitive Linguistics and Cognitive Language Processing. In *Proceedings of the 14th International Conference on Cognitive Modeling (ICCM)*.

Peter Lindes and John Laird. 2017a. Cognitive Modeling Approaches to Language Comprehension Using Construction Grammar. In 2017 AAAI Spring Symposium Series, Computational Construction Grammar and Natural Language Understanding. https://aaai.org/ocs/index.php/SSS/SSS17/paper/view/15285.

Peter Lindes and John E Laird. 2017b. Ambiguity Resolution in a Cognitive Model of Language Comprehension. To be published in *Proceedings of the 15th International Conference on Cognitive Modeling (ICCM 2017)*. In press.

Aaron Mininger and John E Laird. 2016. Interactively Learning Strategies for Handling References to Unseen or Unknown Objects. *Adv. Cogn. Syst 4*.

Shiwali Mohan and John E Laird. 2014. Learning Goal-Oriented Hierarchical Tasks from Situated interactive instruction. In *AAAI*. pages 387–394.

Shiwali Mohan, Aaron Mininger, James Kirk, and John Laird. 2012. Acquiring Grounded Representations of Words with Situated Interactive Instruction. *Advances in Cognitive Systems* 2:113–130.

Luc Steels and Manfred Hild, editors. 2012. Language Grounding in Robots. Springer, New York. https://doi.org/10.1007/978-1-4614-3064-3.

Stefanie Tellex, Ross Knepper, Adrian Li, Daniela Rus, and Nicholas Roy. 2014. Asking for help using inverse semantics. In *Proceedings of Robotics: Science and Systems*. Berkeley, USA. https://doi.org/10.15607/RSS.2014.X.024.

Sida I. Wang, Percy Liang, and Christopher D. Manning. 2016. Learning Language Games Through Interaction. arXiv preprint **arXiv:1606.02447**.

Luke S. Zettlemoyer and Michael Collins. 2007. *Proceedings of the 2007 Joint Conference on Empirical Methods in Natural Language Processing and Computational Natural Language Learning*, 768-687.

# Learning how to learn: an adaptive dialogue agent for incrementally learning visually grounded word meanings


**Yanchao Yu**
Interaction Lab
Heriot-Watt University
`y.yu@hw.ac.uk`

**Arash Eshghi**
Interaction Lab
Heriot-Watt University
`a.eshghi@hw.ac.uk`

**Oliver Lemon**
Interaction Lab
Heriot-Watt University
`o.lemon@hw.ac.uk`



## Abstract

We present an optimised multi-modal dialogue agent for interactive learning of visually grounded word meanings from a human tutor, trained on real human-human tutoring data. Within a life-long interactive learning period, the agent, trained using Reinforcement Learning (RL), must be able to handle natural conversations with human users, and achieve good learning performance (i.e. accuracy) while minimising human effort in the learning process. We train and evaluate this system in interaction with a simulated human tutor, which is built on the BURCHAK corpus – a Human-Human Dialogue dataset for the visual learning task. The results show that: 1) The learned policy can coherently interact with the simulated user to achieve the goal of the task (i.e. learning visual attributes of objects, e.g. colour and shape); and 2) it finds a better trade-off between classifier accuracy and tutoring costs than hand-crafted rule-based policies, including ones with dynamic policies.


## 1 Introduction

As intelligent systems/robots are brought out of the laboratory and into the physical world, they must become capable of natural everyday conversation with their human users about their physical surroundings. Among other competencies, this involves the ability to learn and adapt mappings between words, phrases, and sentences in Natural Language (NL) and perceptual aspects of the external environment – this is widely known as *the grounding problem*.

The grounding problem can be categorised into two distinct, but interdependent types of problem: 1) agent as a second-language learner: the

| Image | Human-Human Dialogue |
|---|---|
|  | T(utor): do you know this object? |
|  | L(earner): a suzuli ... wait no ... sako wakaki? |
|  | T: the color is right, but the shape is not. |
|  | L: oh, okay, so? |
|  | T: a burchak, burchak, sako burchak. |
|  | L: cool, got it. |
|  | L: what is this? |
|  | T: en ... a aylana suzili. |
|  | L: is aylana for color? |
|  | T: no, it's a shape. |
|  | L: so it is an suzili aylana, right? |
|  | T: yes. |

Figure 1: Human-Human Example Dialogues in the BURCHAK Corpus (Yu et al., 2017) ('sako' for 'red', 'burchak' for 'square', 'suzuli' for 'green', 'aylana' for 'circle', 'wakaki' for 'triangle')

agent needs to learn to ground (map) NL symbols onto their existing perceptual and lexical knowledge (e.g. a dictionary of pre-trained classifiers) as in e.g. Silberer and Lapata (2014); Thomason et al. (2016); Kollar et al. (2013); Matuszek et al. (2014); and 2) the agent as a child: without any prior knowledge of perceptual categories, the agent must learn both the perceptual categories themselves and also how NL expressions map to these (Skocaj et al., 2016; Yu et al., 2016c). Here, we concentrate on the latter scenario, where a system learns to identify and describe visual attributes (colour and shape in this case) through interaction with human tutors, incrementally, over time.

Previous work has approached the grounding problem using a variety of resources and approaches, for instance, either using annotated visual datasets (Silberer and Lapata, 2014; Socher et al., 2014; Naim et al., 2015; Al-Omari et al., 2016; Tellex et al., 2014; Matuszek et al., 2012, 2014), or through interactions with other agents or real humans (Kollar et al., 2013; Tellex et al., 2013; Thomason et al., 2015, 2016; Skocaj et al., 2016; Yu et al., 2016c), where feedback from other




*Proceedings of the First Workshop on Language Grounding for Robotics*, pages 10–19,
Vancouver, Canada, July 30 - August 4, 2017. ©2017 Association for Computational Linguistics

agents is used to learn new concepts.

However, most of these systems, which ground NL symbols through interaction have two common, important drawbacks: 1) in order to achieve better performance (i.e. high accuracy), these systems require a high level of human involvement – they always request feedback from human users, which might affect the quality of human answers and decrease the overall user experience in a lifelong learning task; 2) Most of these approaches are not built/trained based on real human-human conversations, and therefore can't handle them. Natural human dialogue is generally more messy than either machine-machine or human-machine dialogue, containing natural dialogue phenomena that are notoriously difficult to capture, e.g. *self- corrections, repetitions and restarts, pauses, fillers, interruptions, and continuations* (Purver et al., 2009; Hough, 2015). Furthermore, they often exhibit much more variation than in their synthetic counterparts (see dialogue examples in Fig. 1).

In order to cope with the first problem, recent prior work (Yu et al., 2016b,c) has built multimodal dialogue systems to investigate the effects of different dialogue strategies and capabilities on the overall learning performance. Their results have shown that, in order to achieve a good trade-off between learning performance and human involvement, the agent must be able to take initiative in dialogues, take into account uncertainty of its predictions, as well as cope with natural human conversation in the learning process. However, their systems are built based on hand-crafted, synthetic dialogue examples rather than real human-human dialogues.

In this paper, we extend this work to introduce an adaptive visual-attribute learning agent trained using Reinforcement Learning (RL). The agent, trained with a multi-objective policy, is capable not only of properly learning novel visual objects/attributes through interaction with human tutors, but also of efficiently minimising human involvement in the learning process. It can achieve equivalent/comparable learning performance (i.e. accuracy) to a fully-supervised system, but with less tutoring effort. The dialogue control policy is trained on the BURCHAK Human-Human Dialogue dataset (Yu et al., 2017), consisting of conversations between a human 'tutor' and a human 'learner' on a visual attribute learning task. The dataset includes a wide range of natural, *incre-*

*mental* dialogue phenomena (such as overlapping turns, self-correction, repetition, fillers, and continuations), as well as considerable variation in the dialogue strategies used by the tutors and the learners.

Here we compare the new optimised learning agent to rule-based agents with and without adaptive confidence thresholds (see section 3.2.1). The results show that the RL-based learning agent outperforms the rule-based systems by finding a better trade-off between learning performance and the tutoring effort/cost.

## 2   Related Work

In this section, we review some of the work that has addressed the language grounding problem generally. The problem of grounding NL in perception has received very considerable attention in the computational literature recently. On the one hand, there is work that only addresses the grounding problem implicitly/indirectly: in this category of work is the large literature on image and video captioning systems that learn to associate an image or video with NL descriptions (Silberer and Lapata, 2014; Bruni et al., 2014; Socher et al., 2014; Naim et al., 2015; Al-Omari et al., 2016). This line of work uses various forms of neural modeling to discover the association between information from multiple modalities. This often works by projecting vector representations from the different modalities (e.g. vision and language) into the same space in order to retrieve one from the other. Importantly, these models are holistic in that they learn to use NL symbols in specific tasks without any explicit encoding of the symbol-perception link, so that this relationship remains implicit and indirect.

On the other hand, other models assume a much more explicit connection between symbols (either words or predicate symbols of some logical language) and perceptions (Kennington and Schlangen, 2015; Yu et al., 2016c; Skocaj et al., 2016; Dobnik et al., 2014; Matuszek et al., 2014). In this line of work, representations are both compositional and transparent, with their constituent atomic parts grounded individually in perceptual classifiers. Our work in this paper is in the spirit of the latter.

Another dimension along which work on grounding can be compared is whether groundings are learned offline (e.g. from images or videos annotated with descriptions or definite reference expressions as in (Kennington and Schlangen, 2015; Socher et al., 2014)) or from live interaction as in, e.g. (Skocaj et al., 2016; Yu et al., 2015, 2016c; Das et al., 2017, 2016; de Vries et al., 2016; Thomason et al., 2015, 2016; Tellex et al., 2013). The latter, which we do here, is clearly more appropriate for multimodal systems or robots that are expected to continuously, and incrementally learn from the environment and their users.

Multi-modal, interactive systems that involve grounded language are either: (1) *rule-based* as in e.g. Skocaj et al. (2016); Yu et al. (2016b); Thomason et al. (2015, 2016); Tellex et al. (2013); Schlangen (2016): in such systems, the dialogue control policy is hand-crafted, and therefore these systems are *static*, cannot adapt, and are less robust; or (2) *optimised* as in e.g. Yu et al. (2016c); Mohan et al. (2012); Whitney et al. (fcmng); Das et al. (2017): in contrast such systems are learned from data, and live interaction with their users; they can thus *adapt* their behaviour dynamically not only to particular dialogue histories, but also to the specific information they have in another modality (e.g. a particular image or video).

Ideally, such interactive systems ought to be able to handle natural, spontaneous human dialogue. However, most work on interactive language grounding learn their systems from synthetic, hand-made dialogues or simulations which lack both in variation and the kinds of dialogue phenomena that occur in everyday conversation; they thus lead to systems which are not robust and cannot handle everyday conversation (Yu et al., 2016c; Skocaj et al., 2016; Yu et al., 2016a). In this paper, we try to change this by training an adaptive learning agent from *human-human dialogues in a visual attribute learning task.*

Given the above, what we achieve here is: we have trained an adaptive attribute-learning dialogue policy from realistic human-human conversations that learns to optimise the trade-off between a learning/grounding performance (*Accuracy*) and costs form human tutors,in effect doing a form of active learning.

## 3  Learning How to Learn Visual Attributes: an Adaptive Dialogue Agent

We build a multimodal and teachable system that supports a visual attribute (e.g. colour and shape) learning process through natural conversational interaction with human tutors (see Fig. 1 for example dialogues), where the tutor and the learner interactively exchange information about the visual attributes of an object they can both see. Here we use Reinforcement Learning for policy optimisation for the learner side (see below Section 3.2). The tutor side is simulated in a data-driven fashion using human-human dialogue data (see below, Sections 4 & 5.2).

### 3.1  Overall System Architecture

The system architecture loosely follows that of Yu et al. (2016c), and employs two core modules:

**Vision Module**   produces visual attribute predictions, using two base feature categories, i.e. the HSV colour space for colour attributes, and a 'bag of visual words' (i.e. PHOW descriptors) for the object shapes/class. It consists of a set of binary classifiers - Logistic Regression SVM classifiers with Stochastic Gradient Descent (SGD) (Zhang, 2004) – to incrementally learn attribute predictions. The visual classifiers ground visual attribute words such as 'red', 'circle' etc. that appear as parameters of the Dialogue Acts used in the system.

**Dialogue Module**   that implements a dialogue system with a classical architecture, composed of Dialogue Management (DM), Natural Language Understanding (NLU) and Generation (NLG) components. The components interact via Dialogue Act representations (e.g. `inform(color=red),ask(shape)`). It is these action representations that are grounded in the visual classifiers that reside in the vision module. The DM relies on an adaptive policy that is learned using RL. The policy is trained to: 1) handle natural interactions with humans and to produce coherent dialogues; and 2) optimise the trade-off between accuracy of visual classifiers and the cost of the dialogue to the tutor.

### 3.2  Adaptive Learning Agent with Hierarchical MDP

Given the visual attribute learning task, the smart agent must learn novel visual objects/attributes as accurately as possible through natural interactions with real humans, but meanwhile it should attempt to minimise the human involvement as much as possible in this life-long learning process. We formulate this interactive learning task into two sub-tasks, which are trained using Reinforcement Learning with a hierarchical Markov

Decision Process (MDP), consisting of two inter-dependent MDPs (sections 3.2.1 and 3.2.2):

### 3.2.1 Adaptive Confidence Threshold

Following previous work (Yu et al., 2016c), we also here use a positive confidence threshold: this is a threshold which determines when the agent believes its own predictions. This threshold plays an essential role in achieving the trade-off between the learning performance and the tutoring cost, since the agent's behaviour, e.g. whether to seek feedback from the tutor, is dependent on this threshold. A form of *active learning* is taking place: the learner only asks a question about an attribute if it isn't confident enough already about that attribute.

Here, we learn an adaptive strategy that aims at maximising the overall learning performance simultaneously, by properly adjusting the positive confidence threshold in the range of 0.65 to 0.95. We train the optimization using a RL library – Burlap (MacGlashan, 2015) as follows, in detail:

**State Space** The adaptive-threshold MDP initialises a 3-dimensional state space defined by $Num_{Instance}$, $Threshold_{cur}$, and $deltaAcc$, where $Num_{Instance}$ represents how many visual objects/images have been seen (the number of instances will be clustered into 50 bins, each bin contains 10 visual instances); $Threshold_{cur}$ represents the positive threshold the agent is currently applying; and $deltaAcc$ represents, after seeing each 10 instances, whether the classifier accuracy increases, decreases or keep constant comparing to the previous bin. The $deltaAcc$ is configured into three levels, (see Eq.1)

$$deltaAcc = \begin{cases} 1, & \text{if } \Delta Acc > 0 \\ 0, & \text{else if } \Delta Acc = 0 \\ -1, & \text{otherwise} \end{cases} \quad (1)$$

**Action Selection** the actions were either to increase or decrease the confidence threshold by 0.05, or keep it the same.

**Reward signal** The reward function for the learning tasks is given by a local function $R_{local}$. This local reward signal was directly proportional to the agents delta accuracy over the previous Learning Step (10 training instances, see above). The single training episode will be terminated once the agent goes through 500 instances.

### 3.2.2 Natural Interaction

The second sub-task aims at learning an optimised dialogue strategy that allows the system to achieve the learning task (i.e. learn new visual attributes) through natural, human-like conversations.

**State Space** The dialogue agent initialises a 4-dimensional state space defined by ($C_{state}$, $S_{state}$, $preDAts$, $preContext$), where $C_{state}$ and $S_{state}$ are the status of visual predictions for the colour and shape attributes respectively (where the status is determined by the prediction score ($conf.$) and the adaptive confidence threshold ($posThd.$) described above (see Eq.2)), the $preDAts$ represents the previous dialogue actions from the tutor response, and the $preContext$ represents which attribute categories (e.g. colour, shape or both) were talked about in the context history.

$$State = \begin{cases} 2, & \text{if } conf. \geq posThd \\ 1, & \text{else if } 0.5 < conf. < posThd. \\ 0, & \text{otherwise} \end{cases}$$
$$(2)$$

i.e. $C_{state}$ or $S_{state}$ will be updated to 2 also when the related knowledge has been provided by the tutor.

**Action Selection** The actions were chosen based on the statistics of the dialog action frequency occurred from the BURCHAK corpus, including *question-asking(for WH questions or polar questions)*, *inform*, *acknowledgment*, as well as *listening*. These actions can be applied for either specific single attribute or both. The action of *inform* can be separated into two sub-actions according to whether the prediction score is greater than 0.5 (i.e. *polar question*) or not (i.e. *doNotKnow*).

**Reward signal** The reward function for the learning tasks is given by a global function $R_{global}$ (see Eq.3). The dialogue will be terminated when both colour and shape knowledge are either taught by human tutors or known with high confidence scores.

$$R_{global} = 10 - C_{ost} - penal.; \quad (3)$$

where $C_{ost}$ represents the cumulative cost by the tutor (see more details about this setup in Section 5.1) in a single dialogue, and $penal.$ penalizes all performed actions which cannot respond to the user properly.

| Dialogue Capability | Speaker | Annotation Tag |
| --- | --- | --- |
| Listen | Tutor/Learner | Listen() |
| Inform | Tutor/Leaner | Inform(colour:sako&shape:burchak) |
| Question_asking | Tutor/Leaner | Ask(colour), Ask(shape), Ask(colour&shape) |
| Question-answering | Tutor/Leaner | Inform(colour:sako), Polar(shape:burchak) |
| Acknowledgement | Tutor/Learner | Ack(), Ack(colour) |
| Rejection | Tutor | Reject(), Reject(shape) |
| Focus | Tutor | Focus(colour), Focus(shape) |
| Clarification | Tutor | CLr() |
| Clarification-request | Learner | CLrRequest() |
| Help-offer | Tutor | Help() |
| Help-request | Learner | HelpRequest() |
| Checking | Tutor | Check() |
| Repetition-request | Tutor | Repeat() |
| Retry-request | Tutor | Retry() |

Table 1: List of Dialogue Capabilities/Actions and Corresponding Annotations in the Corpus

i.e. we applied the **SARSA algorithm** (Sutton and Barto, 1998) for learning the multi-MDP learning agent with each episode defined as a complete dialogue for an object. It was configured with a $\xi-$Greedy exploration rate of 0.2 and a discount factor of 1.

## 4 Human-Human Dialogue Corpus: BURCHAK

BURCHAK (Yu et al., 2017) is a freely available Human-Human Dialogue dataset consisting of 177 dialogues between real human users on the task of interactively learning visual attributes.

**The DiET experimental toolkit** These dialogue were collected using a new *incremental variation* of the DiET chat-tool developed by (Healey et al., 2003; Mills and Healey, submitted), which allows two or more participants to communicate in a shared chat window. It supports live, fine-grained and highly local experimental manipulations of ongoing human-human conversation (see e.g. (Eshghi and Healey, 2015)). The chat-tool is designed to support, elicit, and record at a fine-grained level, dialogues that resemble face-to-face dialogue in that turns are: (1) constructed and displayed incrementally as they are typed; (2) transient; (3) potentially overlapping; (4) not editable, i.e. deletion is not permitted.

**Task** The learning/tutoring task given to the participants involves a pair of participants who talk about visual attributes (e.g. colour and shape) through a series of visual objects. The overall goal of this task is for the learner to discover groundings between visual attribute words and aspects in the physical world through interaction. However, since humans have already known all groundings, such as "red" and "square", the task is assumed in a second-language learning scenario, where each visual attribute, instead of standard English words, is assigned to a new unknown word in a made-up language (see examples in Fig. 1). (see more details in (Yu et al., 2017))

**Dialogue Phenomena** As the chat-tool is designed to resemble face-to-face dialogue, the most important challenge of this BURCHAK is that it refers to a wide range of natural, incremental dialogue phenomena, such as overlapping, self-correction and repetition, filler as well as continuation (Fig. 1). On the other hand, BURCHAK, which focuses on the visual attribute learning task, offers a list of interesting task-oriented dialogue strategies (e.g. initiative, context-dependency and knowledge-acquisition) and capabilities, such as inform, question-asking and answering, listen (no act), as well as acknowledgement and rejection. Each dialogue action contains a huge variations in the realistic conversation. All dialogue actions are tagged in the dataset (as shown in Table 1).

i.e. we have trained and evaluated the optimised learning agents on the cleaned-up version of this corpus, in which spelling mistakes, emoticons, as well as some snippets of conversations where the participant misunderstood the task have been corrected or removed.

## 5  Experiment Setup

In this section, we follow previous work (Yu et al., 2016c) to compare the trained RL-based learning agent with a rule-based system with the best performance (i.e. an agent which takes the initiative in dialogues, takes into account its changing confidence about its predictions, and is also able to process natural, human-like dialogues) from previous work. Instead of using hand-crafted dialogue examples as before, both the RL-based system and the rule-based system are trained/developed against a simulated user, itself trained from the BURCHAK dialogue data set as above. For learning simple visual attributes (e.g. *"red"* and *"square"*), we use the same hand-made visual object dataset from Yu et al. (2016c).

In order to further investigate the effects of the optimised adaptive confidence threshold on the learning performance, we build the rule-based system under three different settings, i.e. with a constant threshold (0.95) (see *blue* curve in Fig. 2), with a hand-crafted adaptive threshold which drops by 0.05 after each 10 instances (*grey* curve in Fig. 2), and with a hand-crafted adaptive threshold which drops by 0.01 after each 10 instances (*orange* curve in Fig. 2).

### 5.1  Evaluation Metrics

To compare the optimised and the rule-based learning agents, and also further investigate how the adaptive threshold affect the learning process, we follows the evaluate metrics from the previous work (see (Yu et al., 2016c)) considering both the cost to the tutor and the accuracy of the learned meanings, i.e. the classifiers that ground our colour and shape concepts.

**Cost**   The cost measure reflects the effort needed by a human tutor in interacting with the system. Skocaj et. al. (2009) point out that a comprehensive teachable system should learn as autonomously as possible, rather than involving the human tutor too frequently. There are several possible costs that the tutor might incur: $C_{inf}$ refers to the cost (*i.e.* $5\,points$) of the tutor providing information on a single attribute concept (e.g. "this is red" or "this is a square"); $C_{ack}$ is the cost (*i.e.* 0.5) for a simple confirmation (like "yes", "right") or rejection (such as "no"); $C_{crt}$ is the cost of correction for a single concept (e.g. "no, it is blue" or "no, it is a circle"). We associate a higher cost (*i.e.* 5) with correction of statements than that

of polar questions. This is to penalise the learning agent when it confidently makes a false statement – thereby incorporating an aspect of trust in the metric (humans will not trust systems which confidently make false statements).
i.e. differently to the previous evaluation metrics, we do not take into account the costs of parsing and producing utterances

**Learning Performance**   As mentioned above, an efficient learner dialogue policy should consider both classification accuracy and tutor effort (Cost). We thus define an integrated measure – the *Overall Performance Ratio* ($R_{perf}$) – that we use to compare the learner's overall performance across the different conditions:

$$R_{perf} = \frac{\Delta Acc}{C_{tutor}}$$

i.e. the increase in accuracy per unit of the cost, or equivalently the gradient of the curve in Fig. 2c. We seek dialogue strategies that maximise this.

### 5.2  User Simulation

In order to train and evaluate these learning agents, we build an user simulation using a generic n-gram framework (see (Yu et al., 2017)) on the BURCHAK corpus. This user framework takes as input the sequence of N most recent words in the dialogue, as well as some optional additional conditions, and then outputs the next user response on multiple levels as required, e.g. full utterance, a sequence of dialogue actions, or even a sequence of single word outputs for incremental dialogue. Differently to other existing user simulations, this framework aims at not only resembling user strategies and capabilities in realistic conversations, but also at simulating incremental dialogue phenomena, e.g. self-repair and repetition, and pauses, as well as fillers. In this paper, we created an action-based user model that predict the next user response in a sequence of dialogue actions. The simulator then produces a full utterance by following the statistics of utterance templates for each predicted action.

### 5.3  Results

Table 2 shows example interactions between the learned RL agent and the simulated tutor on the learning task. The dialogue agent learned to take the initiative and constantly produces coherent conversations through the learning process.

| **Dialogue Example (a)** |
| --- |
| T: what is this object called? |
| L: a red square? |
| T: the shape is correct, but the colour is wrong. |
| L: so what colour is this? |
| T: green. |
| L: okay, get it. |

| **Dialogue Example (b)** |
| --- |
| L: blue? |
| T: yes, blue is for the colour. and shape? |
| L: sorry, i don't know the shape. |
| T: the shape is circle. |
| L: okay, got it. |

Table 2: User Simulation Examples for *(a) Tutor takes the initiative (b) Learner takes the initiative*

Fig. 2a and 2b plot the progression of average Accuracy and (cumulative) Tutoring Cost for each of the 4 learning agents in our experiment, as the system interacts over time with the tutor about each of the 500 training instances.

As noted in passing, the vertical axes in these graphs are based on averages across the 20 folds - recall that for Accuracy the system was tested, in each fold, at every learning step, i.e. after every 10 training instances.

Fig. 2c, on the other hand, plots Accuracy against Tutoring Cost directly. Note that it is to be expected that the curves should not terminate in the same place on the x-axis since the different conditions incur different total costs for the tutor across the 500 training instances. The gradient of this curve corresponds to *increase in Accuracy per unit of the Tutoring Cost*. It is the gradient of the line drawn from the beginning to the end of each curve ($tan(\beta)$ on Fig. 2c) that constitutes our main evaluation measure of the system's overall performance in each condition, and it is this measure for which we report statistical significance results: there are significant differences in accuracy between the RL-based policy and two rule-based policies with the hand-crafted threshold ($p < 0.01$ for both). The RL-based policy shows significantly less tutoring cost than the rule-based system with a constant threshold ($p < 0.01$). The mean gradient of the yellow, RL curve is actually slightly higher than the constant-threshold policy blue curve - discussed below.

## 5.4  Discussion

**Accuracy**  As can be seen in Fig. 2a, the rule-based system with a constant threshold (0.95) shows the fastest increase in accuracy and finally reaches around 0.87 at the end of the learning process (i.e. after seeing 500 instances) – the blue curve. Both systems with a hand-crafted adaptive threshold, with an incremental decrease of 0.01 (grey curve) and 0.05 (orange curve), have shown an unexpected trend in accuracy across 500 instances, where the orange curve flattens out at about 0.76 after seeing only 50 instances, and the grey curve shows a good increase in the beginning but later drops down to about 0.77 after 150 instances. This is because the thresholds were decreased too fast, so that the agent cannot hear enough feedback (i.e. corrective attribute labels) from tutors to improve its predictions. In contrast to this, the optimised RL-based agent achieves much better accuracy (i.e. about 0.85) by the end of the experiment.

**Tutoring Cost**  As mentioned above, there is a form of *active learning* taking place in the experiment: the agent can only hear feedback from the tutor if it is not confident enough about its own predictions. This also explains the slight decrease in the gradients of the curves (i.e. the cumulative cost for the tutor) (see Fig. 2b) as the agent is exposed to more and more training instances: its subjective confidence about its own predictions increases over time, and thus there is progressively less need for tutoring. In detail, the tutoring cost progresses much more slowly while the system was applying a hand-crafted adaptive threshold (i.e. incrementally decreases by either 0.01 or 0.05 after each bin). This is still because there were not interactions taking place at all once the threshold is lower than a certain value (for instance, 0.65), where the agent might be highly confident on all its predictions. In contrast, the RL-based agent shows a faster progress in the cumulative tutoring cost, but achieves higher accuracy.

**Overall Performance**  Here, we only compare the gradients of the curves between the optimised learning agent (yellow curve) and the rule-based system with a constant threshold (blue curve) in Fig. 2c, because others with the incremental decreased threshold cannot achieve an acceptable learning performance. The agent with an adaptive threshold (yellow) achieves slightly better overall gradient ($tan(\beta_1)$) than the rule-based system

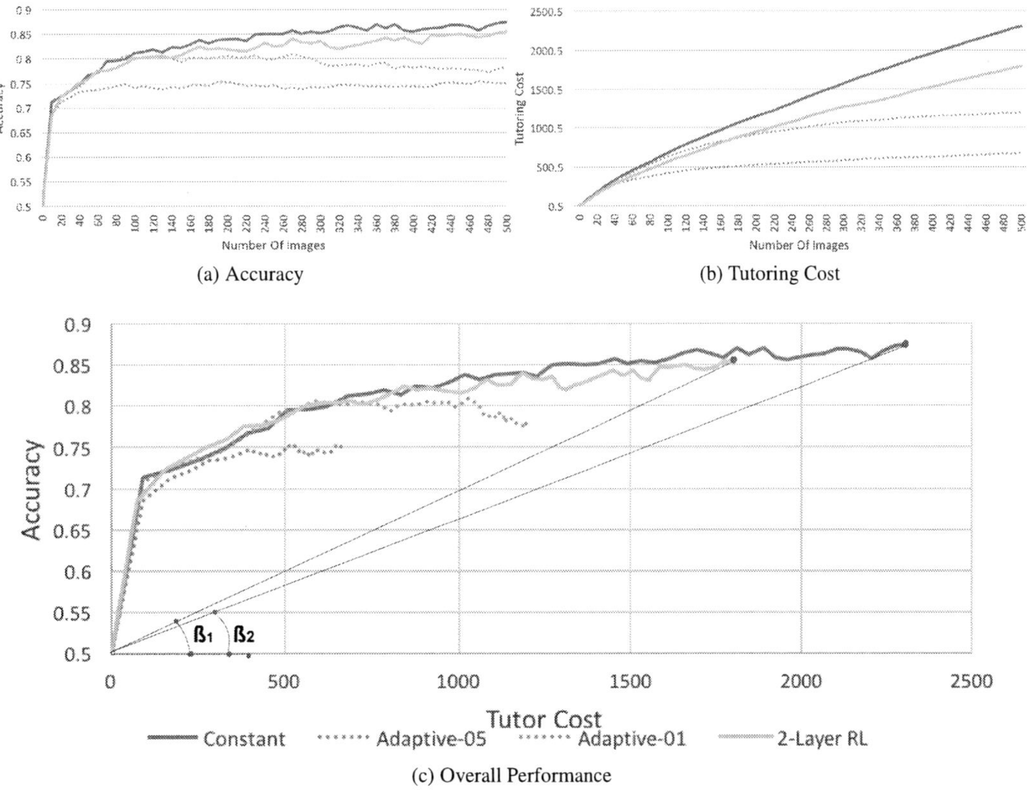

(a) Accuracy       (b) Tutoring Cost



(c) Overall Performance

Figure 2: Evolution of Learning Performance

$(tan(\beta_2))$, it achieves a comparable accuracy and does it faster. We therefore conclude that the optimised learning agent, which finds a better trade-off between the learning accuracy and the tutoring cost, is more desirable.

## 6 Conclusion & Future Work

We have introduced a multi-modal learning agent that can incrementally learn grounded word meanings through interaction with human tutors over time, and deploys an *adaptive* dialogue policy (optimised using Reinforcement Learning). We applied a human-human dialogue dataset (i.e. BUR-CHAK) to train and evaluate the optimised learning agent. We evaluated the system by comparing it to a rule-based system, and results show that: 1) the optimised policy has learned to coherently interact with the simulated user to learn visual attributes of an object (e.g. colour and shape); 2) it achieves comparable learning performance to a rule-based systems, but with less tutoring effort needed from humans.

Ongoing work further applies Reinforcement Learning at the word level to learn a complete, incremental dialogue policy, i.e. which chooses system output at the lexical level (Eshghi and Lemon, 2014; Kalatzis et al., 2016). In addition, instead of acquiring visual concepts for toy objects (i.e. with simple colour and shape), the system has recently been extended to interactively learn about real object classes (e.g. `shampoo`, `apple`). The latest system integrates with a *Self-Organizing Incremental Neural Network* and a deep *Convolutional Neural Network* to learn object classes through interaction with humans incrementally, over time.

## Acknowledgements

This research is supported by the EPSRC, under grant number EP/M01553X/1 (BABBLE project[1]), and by the European Union's Horizon 2020 research and innovation programme under grant agreement No. 688147 (MuMMER project[2]).

---

[1]`https://sites.google.com/site/hwinteractionlab/babble`
[2]`http://mummer-project.eu/`

# References


Muhannad Al-Omari, Eris Chinellato, Yiannis Gatsoulis, David C. Hogg, and Anthony G. Cohn. 2016. Unsupervised grounding of textual descriptions of object features and actions in video. In *Principles of Knowledge Representation and Reasoning: Proceedings of the Fifteenth International Conference, KR 2016, Cape Town, South Africa, April 25-29, 2016.* pages 505–508.

Elia Bruni, Nam-Khanh Tran, and Marco Baroni. 2014. Multimodal distributional semantics. *J. Artif. Intell. Res.(JAIR)* 49(1–47).

Abhishek Das, Satwik Kottur, Khushi Gupta, Avi Singh, Deshraj Yadav, José M. F. Moura, Devi Parikh, and Dhruv Batra. 2016. Visual dialog. *CoRR* abs/1611.08669.

Abhishek Das, Satwik Kottur, José M. F. Moura, Stefan Lee, and Dhruv Batra. 2017. Learning cooperative visual dialog agents with deep reinforcement learning. *CoRR* abs/1703.06585.

Harm de Vries, Florian Strub, Sarath Chandar, Olivier Pietquin, Hugo Larochelle, and Aaron C. Courville. 2016. Guesswhat?! visual object discovery through multi-modal dialogue. *CoRR* abs/1611.08481.

Simon Dobnik, Robin Cooper, and Staffan Larsson. 2014. Type theory with records: a general framework for modelling spatial language. In *Proceedings of The Second Workshop on Action, Perception and Language (APL'2).*

Arash Eshghi and Patrick G. T. Healey. 2015. Collective contexts in conversation: Grounding by proxy. *Cognitive Science* pages 1–26.

Arash Eshghi and Oliver Lemon. 2014. How domain-general can we be? learning incremental dialogue systems without dialogue acts. In *Proceedings of SemDial.*

P. G. T. Healey, Matthew Purver, James King, Jonathan Ginzburg, and Greg Mills. 2003. Experimenting with clarification in dialogue. In *Proceedings of the 25th Annual Meeting of the Cognitive Science Society.* Boston, Massachusetts.

Julian Hough. 2015. *Modelling Incremental Self-Repair Processing in Dialogue.* Ph.D. thesis, Queen Mary University of London.

Dimitrios Kalatzis, Arash Eshghi, and Oliver Lemon. 2016. Bootstrapping incremental dialogue systems: using linguistic knowledge to learn from minimal data. *CoRR* abs/1612.00347. http://arxiv.org/abs/1612.00347.

Casey Kennington and David Schlangen. 2015. Simple learning and compositional application of perceptually grounded word meanings for incremental reference resolution. In *Proceedings of the 53rd Annual Meeting of the Association for Computational Linguistics and the 7th International Joint Conference on Natural Language Processing of the Asian Federation of Natural Language Processing, ACL 2015, July 26-31, 2015, Beijing, China, Volume 1: Long Papers.* pages 292–301.

Thomas Kollar, Jayant Krishnamurthy, and Grant Strimel. 2013. Toward interactive grounded language acqusition. In *Robotics: Science and Systems.*

James MacGlashan. 2015. Burlap http://burlap.cs.brown.edu/.

Cynthia Matuszek, Liefeng Bo, Luke Zettlemoyer, and Dieter Fox. 2014. Learning from unscripted deictic gesture and language for human-robot interactions. In *Proceedings of the Twenty-Eighth AAAI Conference on Artificial Intelligence, July 27 -31, 2014, Québec City, Québec, Canada.* pages 2556–2563.

Cynthia Matuszek, Nicholas FitzGerald, Luke Zettlemoyer, Liefeng Bo, and Dieter Fox. 2012. A joint model of language and perception for grounded attribute learning. In *Proc. of the 2012 International Conference on Machine Learning.* Edinburgh, Scotland.

Gregory J. Mills and Patrick G. T. Healey. submitted. The Dialogue Experimentation toolkit. *xx* (?).

Shiwali Mohan, Aaron Mininger, James Kirk, and John E. Laird. 2012. Learning grounded language through situated interactive instruction. In *Robots Learning Interactively from Human Teachers, Papers from the 2012 AAAI Fall Symposium, Arlington, Virginia, USA, November 2-4, 2012.*

Iftekhar Naim, Young Chol Song, Qiguang Liu, Liang Huang, Henry A. Kautz, Jiebo Luo, and Daniel Gildea. 2015. Discriminative unsupervised alignment of natural language instructions with corresponding video segments. In *NAACL HLT 2015, The 2015 Conference of the North American Chapter of the Association for Computational Linguistics: Human Language Technologies, Denver, Colorado, USA, May 31 - June 5, 2015.* pages 164–174.

Matthew Purver, Raquel Fernández, Matthew Frampton, and Stanley Peters. 2009. Cascaded lexicalised classifiers for second-person reference resolution. In *Proceedings of the 10th Annual SIGDIAL Meeting on Discourse and Dialogue (SIGDIAL 2009 Conference).* Association for Computational Linguistics, London, UK, pages 306–309. http://www.dcs.qmul.ac.uk/ mpurver/papers/purver-et-al09sigdial-you.pdf.

David Schlangen. 2016. Grounding, justification, adaptation: Towards machines that mean what they say. In *Proceedings of the 20th Workshop on the Semantics and Pragmatics of Dialogue (JerSem).*

Carina Silberer and Mirella Lapata. 2014. Learning grounded meaning representations with autoencoders. In *Proceedings of the 52nd Annual Meeting of the Association for Computational Linguistics*

*(Volume 1: Long Papers)*. Association for Computational Linguistics, Baltimore, Maryland, volume 1, pages 721–732.

Danijel Skocaj, Alen Vrecko, Marko Mahnic, Miroslav Janícek, Geert-Jan M. Kruijff, Marc Hanheide, Nick Hawes, Jeremy L. Wyatt, Thomas Keller, Kai Zhou, Michael Zillich, and Matej Kristan. 2016. An integrated system for interactive continuous learning of categorical knowledge. *J. Exp. Theor. Artif. Intell.* 28(5):823–848. https://doi.org/10.1080/0952813X.2015.1132268.

Danijel Skočaj, Matej Kristan, and Aleš Leonardis. 2009. Formalization of different learning strategies in a continuous learning framework. In *Proceedings of the Ninth International Conference on Epigenetic Robotics; Modeling Cognitive Development in Robotic Systems*. Lund University Cognitive Studies, pages 153–160.

Richard Socher, Andrej Karpathy, Quoc V Le, Christopher D Manning, and Andrew Y Ng. 2014. Grounded compositional semantics for finding and describing images with sentences. *Transactions of the Association for Computational Linguistics* 2:207–218.

Richard S. Sutton and Andrew G. Barto. 1998. *Reinforcement Learning: an Introduction*. MIT Press.

Stefanie Tellex, Pratiksha Thaker, Joshua Mason Joseph, and Nicholas Roy. 2014. Learning perceptually grounded word meanings from unaligned parallel data. *Machine Learning* 94(2):151–167. https://doi.org/10.1007/s10994-013-5383-2.

Stefanie Tellex, Pratiksha Thakerll, Robin Deitsl, Dimitar Simeonovl, Thomas Kollar, and Nicholas Royl. 2013. Toward information theoretic human-robot dialog. *Robotics: Science and Systems* page 409.

Jesse Thomason, Jivko Sinapov, Maxwell Sevtlik, Peter Stone, and Raymond J. Mooney. 2016. Learning multi-modal grounded linguistic semantics by playing "i spy". In *To Appear: Proceedings of the Twenty-Fifth International Joint Conference on Artificial Intelligence, IJCAI-16, New York City, USA, July 9-15, 2016*.

Jesse Thomason, Shiqi Zhang, Raymond J. Mooney, and Peter Stone. 2015. Learning to interpret natural language commands through human-robot dialog. In *Proceedings of the Twenty-Fourth International Joint Conference on Artificial Intelligence, IJCAI 2015, Buenos Aires, Argentina, July 25-31, 2015*. pages 1923–1929.

David Whitney, Eric Rosen, James MacGlashan, and Lawson and Tellex Stefanie L.S. Wong. fcmng. Reducing errors in object-fetching interactions through social feedback. In *Proceedings of the IEEE International Conference on Robotics and Automation ICRA 2017, May 29 – June 3, 2017, Marina Bay Sands, Singapore*.

Yanchao Yu, Arash Eshghi, and Oliver Lemon. 2015. Comparing attribute classifiers for interactive language grounding. In *Proceedings of the Fourth Workshop on Vision and Language*. Association for Computational Linguistics, Lisbon, Portugal, pages 60–69. http://aclweb.org/anthology/W15-2811.

Yanchao Yu, Arash Eshghi, and Oliver Lemon. 2016a. Comparing dialogue strategies for learning grounded language from human tutors. In *Proceedings of Semdial 2016 (JerSem)*. New Jersey.

Yanchao Yu, Arash Eshghi, and Oliver Lemon. 2016b. Interactively learning visually grounded word meanings from a human tutor. In *Proceedings of the 5th Workshop on Vision and Language, hosted by the 54th Annual Meeting of the Association for Computational Linguistics, VL@ACL 2016, August 12, Berlin, Germany*.

Yanchao Yu, Arash Eshghi, and Oliver Lemon. 2016c. Training an adaptive dialogue policy for interactive learning of visually grounded word meanings. In *Proceedings of the SIGDIAL 2016 Conference, The 17th Annual Meeting of the Special Interest Group on Discourse and Dialogue, 13-15 September 2016, Los Angeles, CA, USA*. pages 339–349.

Yanchao Yu, Arash Eshghi, Gregory Mills, and Oliver Lemon. 2017. *Proceedings of the Sixth Workshop on Vision and Language*, Association for Computational Linguistics, chapter The BURCHAK corpus: a Challenge Data Set for Interactive Learning of Visually Grounded Word Meanings, pages 1–10. http://aclweb.org/anthology/W17-2001.

Tong Zhang. 2004. Solving large scale linear prediction problems using stochastic gradient descent algorithms. In *Proceedings of the twenty-first international conference on Machine learning*. ACM, page 116.

# Guiding Interaction Behaviors for
# Multi-modal Grounded Language Learning

**Jesse Thomason, Jivko Sinapov,** and **Raymond J. Mooney**
Department of Computer Science, University of Texas at Austin
Austin, TX 78712, USA
{jesse, jsinapov, mooney}@cs.utexas.edu

## Abstract

Multi-modal grounded language learning connects language predicates to physical properties of objects in the world. Sensing with multiple modalities, such as audio, haptics, and visual colors and shapes while performing interaction behaviors like lifting, dropping, and looking on objects enables a robot to ground non-visual predicates like "empty" as well as visual predicates like "red". Previous work has established that grounding in multi-modal space improves performance on object retrieval from human descriptions. In this work, we gather behavior annotations from humans and demonstrate that these improve language grounding performance by allowing a system to focus on relevant behaviors for words like "white" or "half-full" that can be understood by looking or lifting, respectively. We also explore adding modality annotations (whether to focus on audio or haptics when performing a behavior), which improves performance, and sharing information between linguistically related predicates (if "green" is a color, "white" is a color), which improves grounding recall but at the cost of precision.

## 1 Introduction

Connecting human language predicates like "red" and "heavy" to machine perception is part of the *symbol grounding problem* (Harnad, 1990), approached in machine learning as *grounded language learning*. For many years, grounded language learning has been performed primarily in visual space (Roy and Pentland, 2002; Liu et al., 2014; Malinowski and Fritz, 2014; Mohan et al.,

2013; Sun et al., 2013; Dindo and Zambuto, 2010; Vogel et al., 2010). Recently, researchers have explored grounding in audio (Kiela and Clark, 2015), haptic (Alomari et al., 2017), and multi-modal (Thomason et al., 2016) spaces. Multi-modal grounding allows a system to connect language predicates like "rattles", "empty", and "red" to their audio, haptic, and color signatures, respectively.

Past work has used human-robot interaction to gather language predicate labels for objects in the world (Parde et al., 2015; Thomason et al., 2016). Using only human-robot interaction to gather labels, a system needs to learn effectively from only a few examples. Gathering audio and haptic perceptual information requires doing more than looking at each object. In past work, multiple interaction behaviors are used to explore objects and add this audio and haptic information (Sinapov et al., 2014).

In this work, we gather annotations on what exploratory behaviors humans would perform to determine whether language predicates apply to a novel object. A robot could gather such information by asking human users which action would best allow it to test a particular property, e.g. "To tell whether something is 'heavy' should I look at it or pick it up?" Figure 1 shows some of the behaviors used by our robot in previous work to perceive objects and their properties. In this paper, we show that providing a language grounding system with behavior annotation information improves classification performance on whether predicates apply to objects, despite having sparse predicate-object labels.

We additionally explore adding modality annotations (e.g. is a predicate more auditory or more haptic), drawing on previous work in psychology that gathered modality norms for many words (Lynott and Connell, 2009). Finally, we explore using



*Proceedings of the First Workshop on Language Grounding for Robotics*, pages 20–24,
Vancouver, Canada, July 30 - August 4, 2017. ©2017 Association for Computational Linguistics

grasp     lift     lower

drop     press     push

Figure 1: Behaviors the robot used to explore objects. In addition, the *hold* behavior (not shown) was performed after the *lift* behavior by holding the object in place for half a second. The *look* behavior (not shown) was also performed for all objects.

word embeddings to help with infrequently seen predicates by sharing information with more common ones (e.g. if "thin" is common and "narrow" is rare, we can exploit the fact that they are linguistically related to help understand the latter).

## 2 Dataset and Methodology

Previous work provides sparse annotations of 32 household objects (Figure 2) with language predicates derived during an interactive "I Spy" game with human users (Thomason et al., 2016). Each predicate $p \in P$ from that work is associated with objects as applying or not applying, based on dialog with human users. For example, predicate "red" applies to several objects and not to others, but for many objects its label is not explicitly known. Objects are represented by features gathered during several interaction behaviors (Figure 1) as detailed in past work (Sinapov et al., 2016). In this work, we focus on improving the language grounding performance of multimodal classifiers that predict whether each predicate $p \in P$ applies to each object $o \in O$.

In previous work, decisions about a predicate and an object are made for each sensorimotor context (a combination of a behavior and sensory modality) with an SVM using the feature space for that context (Thomason et al., 2016). A summary of sensorimotor contexts is given in Table 1.

Figure 2: Objects explored via interaction behaviors and for which we have sparse predicate annotations.

| Behaviors | Modalities |
|---|---|
| look | color, fpfh |
| drop, grasp, hold, lift lower, press, push | audio, haptics |

Table 1: The *contexts* (combinations of robot behavior and perceptual modality) we use for multimodal language grounding. The *color* modality is color histograms, *fpfh* is fast-point feature histograms, *audio* is fast Fourier transform frequency bins, and *haptics* is averages over robot arm joint forces (detailed in (Sinapov et al., 2016)).

For example, a classifier is trained from the positive and negative object examples for "red" in *look/color* space as well as in the less relevant *drop/audio* space. These decisions are then averaged together, each weighted by its Cohen's-$\kappa$ agreement with human labels using leave-one-out cross validation on the training data. In this way, the *look/color* space for "red" is expected to have high $\kappa$ and a large influence on the decision, while *drop/audio* would have low $\kappa$ and not influence the decision much.

The decision $d(p, o) \in [-1, 1]$ for predicate $p$ and object $o$ is defined as:

$$d(p, o) = \sum_{c \in C} \kappa_{p,c} G_{p,c}(o), \qquad (1)$$

for $G_{p,c}$ a supervised grounding classifier trained on labeled objects for predicate $p$ in the feature space of sensorimotor context $c$ that returns in $\{-1, 1\}$ with $\kappa_{p,c}$ its agreement with human labels. If $d(p, o) \leq 0$, we say $p$ does not apply to $o$, else that it does. We use SVMs with linear kernels as grounding classifiers.

We extend the weighting scheme between sensorimotor SVMs to include behavior information. For each predicate derived from the "I Spy" game

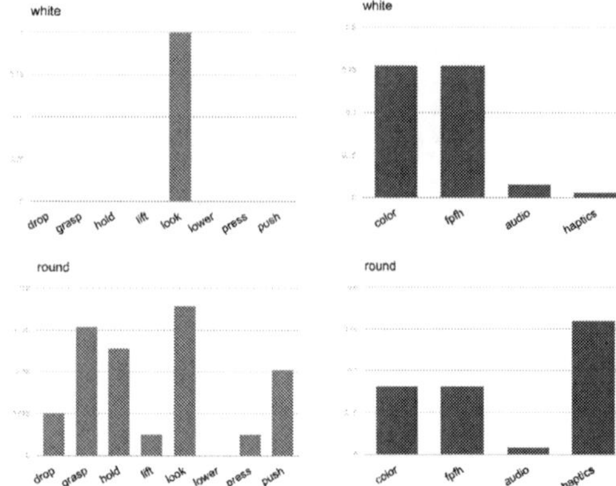

Figure 3: The distribution over annotator-chosen behaviors (left) gathered in this work, as well as the distribution over modality norms (right) derived from previous work (Lynott and Connell, 2009), for the predicates "white" and "round". The *fpfh* modality is fast-point feature histograms.

in previous work, we gather relevant behaviors from human annotators. Annotators were asked to mark which among the 8 exploratory behaviors (Table 1) they would engage in to determine whether a given predicate applied to a novel object. Annotators could mark as many behaviors as they wanted for each predicate, but were required to choose at least one.

We gathered annotations from 14 people, then discarded the annotations from those whose average $\kappa$ agreement with all other annotators was less than $0.4$ (the poor-fair agreement range). This left us with 8 annotators whose average $\kappa = .475$ (moderate agreement). We release the full set of gathered annotations on 81 perceptual predicates across 8 behaviors as a corpus for community use.[1]

Then, for each $p \in P$, we induce a distribution over behaviors $b \in B$ based on the ratio of annotators that marked that behavior relevant, such that $\sum_{b \in B} A_{p,b}^B = 1$, with $A_{p,b}^B$ equal to the proportion of annotators who marked behavior $b$ relevant for understanding predicate $p$. Some predicates, like "white", have single behavior distributions. For other predicates, like "metal", annotators chose more complex combinations of behaviors. Figure 3 (Left) gives some examples of behavior distributions from our annotations.

The decision $d_B(p, o)$ considering behavior annotations is calculated as

$$d_B(p, o) = \sum_{c \in C} A_{p,c_b}^B \kappa_{p,c} G_{p,c}(o), \qquad (2)$$

where $c_b$ is the behavior for sensorimotor context $c$.

We also experiment with adding modality annotations (Table 1). In particular, we derive a modality distribution for each $p \in P$ such that $\sum_{m \in M} A_{p,m}^M = 1$ from modality exclusivity norms gathered by past work for auditory, gustatory, haptic, olfactory, and visual modalities (Lynott and Connell, 2009). We ignore gustatory and olfactory modalities, which have no counterpart in our sensorimotor contexts, and create $A_{p,m}^M$ scores from the auditory, haptic, and visual modality norm means. The visual modality norm is split evenly between relevance scores $A_{p,\text{color}}^M$ and $A_{p,\text{fpfh}}^M$, our visual color and shape modalities.

The decision $d_M(p, o)$ considering modality annotations is calculated as

$$d_M(p, o) = \sum_{c \in C} A_{p,c_b}^M \kappa_{p,c} G_{p,c}(o) \qquad (3)$$

When the predicate $p$ does not appear in the norming dataset from past work[2], a uniform $A_{p,m}^M = 1/|M|$ is used. Figure 3 (Right) gives some examples of modality distributions from these norms.

The data sparsity inherent in language grounding from limited human interaction means some predicates have just a handful of positive and negative examples, while more common predicates may have many. If we have few examples for "narrow" but many for "thin," we can share some information between them. For example, if $\kappa_{\text{thin,grasp/haptic}}$ is high, we should trust the *grasp/haptic* sensorimotor context for "narrow" more than "narrow"'s $\kappa$ estimates alone suggest.

We explore sharing $\kappa$ information between related predicates by calculating their cosine distance in word embedding space by using Word2Vec (Mikolov et al., 2013) vectors derived from Google News.[3] For every pair of predicates $p, q \in P$ with word embedding vectors $v_p, v_q$ we calculate similarity as

$$w(p, q) = \frac{1}{2}(1 + cos(v_p, v_q)), \qquad (4)$$

---



[2] About half the predicates have norming information.
[3] https://github.com/mmihaltz/word2vec-GoogleNews-vectors

| | **p** | **r** | **f1** |
|---|---|---|---|
| **mc** | .282 | .355 | .311 |
| $\kappa$ | .406 | .460 | .422 |
| **B**+$\kappa$ | **.489** | **.489** | **.465** |
| **M**+$\kappa$ | .414 | .466 | .430 |
| **W**+$\kappa$ | .373 | .474 | .412 |

Table 2: Precision (**p**), recall (**r**), and $f1$ (**f1**) of predicate classifiers across weighting schemes. **mc** gives majority class baseline. Weighting schemes consider only validation confidence ($\kappa$, as in previous work), confidence and behavior annotations (**B**+$\kappa$), confidence and modality annotations (**M**+$\kappa$), and confidence and word similarity (**W**+$\kappa$). Note that we show the average per-predicate $f$-measure, not the $f$-measure of the average per-predicate precision and recall.

which falls in $[0, 1]$, and subsequently take a weighted average of $\kappa$ values using these similarities as weights to get decisions $d_W(p, o)$ as

$$d_W(p, o) = \sum_{c \in C} \left( |P|^{-1} \sum_{q \in P} \kappa_{q,c} w(p, q) \right) G_{p,c}(o)$$

(5)

## 3  Experimental Evaluation

We calculated precision, recall, and $f$-measure between human labels and predicate decisions when weighting constituent sensorimotor context classifiers by the schemes described above: kappa confidence only (Eq 1, $\kappa$), adding behavior annotations (Eq 2, **B**+$\kappa$), adding modality annotations (Eq 3, **M**+$\kappa$), and sharing kappas across predicates using word similarity (Eq 5, **W**+$\kappa$).

We calculated these metrics for each predicate[4] and averaged scores across all predicates. We use leave-one-object-out cross validation to obtain performance statistics for each weighting scheme.

Table 2 gives the results for predicates that have at least 3 positive and 3 negative training object examples.[5]

We observe that adding behavior annotations or modality annotations improves performance over

using kappa confidence alone, as was done in past work. Sharing kappa confidences across similar predicates based on their embedding cosine similarity improves recall at the cost of precision.

Adding behavior annotations helps more than adding modality norms, but we gathered behavior annotations for all predicates, while modality annotations were only available for a subset (about half). Adding behavior annotations helped the $f$-measure of predicates like "pink", "green", and "half-full", while adding modality annotations helped with predicates like "round", "white", and "empty".

Sharing confidences through word similarity helped with some predicates, like "round", at the expense of domain-specific meanings of predicates like "water". In the "I Spy" paradigm from which these data were gathered, the authors noted that "water" correlated with object weight because all of their water bottle objects were partially or completely full (Thomason et al., 2016). Thus, in that domain, "water" is synonymous with "heavy". In a less restricted domain, word similarity may add less real world "noise" to the problem.

## 4  Conclusions and Future Work

In this work, we have demonstrated that behavior annotations can improve language grounding for a platform with multiple interaction behaviors and modalities. In the future, we would like to apply this intuition in an embodied dialog agent. If a person asks a service robot to "Get the white cup.", the robot should be able to ask "What should I do to tell if something is 'white'?", a behavior annotation prompt. A human-robot POMDP dialog policy could be learned, as in previous work (Padmakumar et al., 2017), to know when this kind of follow-up question is warranted.

Additionally, we will explore other methods of sharing information between predicates from lexical information. For example, choosing a maximally similar neighboring word, rather than doing a weighted average across all known words, may yield better results (e.g. the best neighbor of "narrow" is "thin", so don't bother considering things like "green" at all).

## Acknowledgments

We thank our anonymous reviewers for their time and insights. This work is supported by a National Science Foundation Graduate Research

---

[4]Decisions were made for each testing object and marked correct or incorrect against human labels that object, if available for the predicate.

[5]The trends are similar when considering all predicates, but the scores and differences in performance are lower due to many predicates having only a single positive or negative example.

Fellowship to the first author, an NSF EAGER grant (IIS-1548567), and an NSF NRI grant (IIS-1637736). A portion of this work has taken place in the Learning Agents Research Group (LARG) at UT Austin. LARG research is supported in part by NSF (CNS-1330072, CNS-1305287, IIS-1637736, IIS-1651089), ONR (21C184-01), AFOSR (FA9550-14-1-0087), Raytheon, Toyota, AT&T, and Lockheed Martin.

# References


Muhannad Alomari, Paul Duckworth, David C. Hogg, and Anthony G. Cohn. 2017. Natural language acquisition and grounding for embodied robotic systems. In *Proceedings of the Thirty-First AAAI Conference on Artificial Intelligence*. pages 4349–4356.

Haris Dindo and Daniele Zambuto. 2010. A probabilistic approach to learning a visually grounded language model through human-robot interaction. In *International Conference on Intelligent Robots and Systems*. IEEE, Taipei, Taiwan, pages 760–796.

S. Harnad. 1990. The symbol grounding problem. *Physica D* 42:335–346.

Douwe Kiela and Stephen Clark. 2015. Multi- and cross-modal semantics beyond vision: Grounding in auditory perception. In *Proceedings of the 2015 Conference on Emperical Methods in Natural Language Processing*. Lisbon, Portugal, pages 2461–2470.

Changson Liu, Lanbo She, Rui Fang, and Joyce Y. Chai. 2014. Probabilistic labeling for efficient referential grounding based on collaborative discourse. In *Proceedings of the 52nd Annual Meeting of the Association for Computational Linguistics*. Baltimore, Maryland, USA, pages 13–18.

Dermot Lynott and Louise Connell. 2009. Modality exclusivity norms for 423 object properties. *Behavior Research Methods* 41(2):558–564.

Mateusz Malinowski and Mario Fritz. 2014. A multi-world approach to question answering about real-world scenes based on uncertain input. In *Proceedings of the 28th Annual Conference on Neural Information Processing Systems*. Montréal, Canada, pages 13–18.

Tomas Mikolov, Ilya Sutskever, Kai Chen, Greg Corrado, and Jeffrey Dean. 2013. Distributed representations of words and phrases and their compositionality. In *Proceedings of the 26th International Conference on Neural Information Processing Systems*. Lake Tahoe, Nevada, pages 3111–3119.

Shiwali Mohan, Aaron H. Mininger, and John E. Laird. 2013. Towards an indexical model of situated language comprehension for real-world cognitive agents. In *Proceedings of the 2nd Annual Conference on Advances in Cognitive Systems*. Baltimore, Maryland, USA.

Aishwarya Padmakumar, Jesse Thomason, and Raymond J. Mooney. 2017. Integrated learning of dialog strategies and semantic parsing. In *Proceedings of the 15th Conference of the European Chapter of the Association for Computational Linguistics*. Valencia, Spain, pages 547–557.

Natalie Parde, Adam Hair, Michalis Papakostas, Konstantinos Tsiakas, Maria Dagioglou, Vangelis Karkaletsis, and Rodney D. Nielsen. 2015. Grounding the meaning of words through vision and interactive gameplay. In *Proceedings of the 24th International Joint Conference on Artificial Intelligence*. Buenos Aires, Argentina, pages 1895–1901.

Deb Roy and Alex Pentland. 2002. Learning words from sights and sounds: a computational model. *Cognitive Science* 26(1):113–146.

Jivko Sinapov, Priyanka Khante, Maxwell Svetlik, and Peter Stone. 2016. Learning to order objects using haptic and proprioceptive exploratory behaviors. In *Proceedings of the 25th International Joint Conference on Artificial Intelligence*.

Jivko Sinapov, Connor Schenck, and Alexander Stoytchev. 2014. Learning relational object categories using behavioral exploration and multimodal perception. In *IEEE International Conference on Robotics and Automation*.

Yuyin Sun, Liefeng Bo, and Dieter Fox. 2013. Attribute based object identification. In *International Conference on Robotics and Automation*. IEEE, Karlsruhe, Germany, pages 2096–2103.

Jesse Thomason, Jivko Sinapov, Maxwell Svetlik, Peter Stone, and Raymond Mooney. 2016. Learning multi-modal grounded linguistic semantics by playing "I spy". In *Proceedings of the 25th International Joint Conference on Artificial Intelligence*. pages 3477–3483.

Adam Vogel, Karthik Raghunathan, and Dan Jurafsky. 2010. Eye spy: Improving vision through dialog. In *Association for the Advancement of Artificial Intelligence*. pages 175–176.

# Structured Learning for Context-aware
# Spoken Language Understanding of Robotic Commands


**Andrea Vanzo**[†] and **Danilo Croce**[‡] and **Roberto Basili**[‡] and **Daniele Nardi**[†]

[†]Sapienza University of Rome

Dept. of Computer, Control and Management Engineering "Antonio Ruberti"

[‡]University of Roma, Tor Vergata

Dept. of Enterprise Engineering

{vanzo,nardi}@dis.uniroma1.it,{croce,basili}@info.uniroma2.it



## Abstract

Service robots are expected to operate in specific environments, where the presence of humans plays a key role. A major feature of such robotics platforms is thus the ability to react to spoken commands. This requires the understanding of the user utterance with an accuracy able to trigger the robot reaction. Such correct interpretation of linguistic exchanges depends on physical, cognitive and language-dependent aspects related to the environment. In this work, we present the empirical evaluation of an adaptive Spoken Language Understanding chain for robotic commands, that explicitly depends on the operational environment during both the learning and recognition stages. The effectiveness of such a context-sensitive command interpretation is tested against an extension of an already existing corpus of commands, that introduced explicit perceptual knowledge: this enabled deeper measures proving that more accurate disambiguation capabilities can be actually obtained.


## 1 Introduction

In recent years, one of the most challenging issues that Service Robotics is facing is the automation of high level and collaborative interactions between humans and robots. In such a robotic context, human language is the most natural way of communication as for its expressiveness and flexibility. However, an effective communication in natural language between humans and robots is challenging mostly for the different cognitive abilities it involves. For a robot to react to a simple command like *"take the mug in the kitchen"*, a number of implicit assumptions should be met. First, at least two entities, a mug and a kitchen, must exist in the environment and the speaker must be aware of such entities. Accordingly, the robot must have access to an inner representation of its world, e.g., an explicit map of the environment. Second, mappings from lexical references to real world entities must be developed or made available. In this respect, the *Grounding* process (Harnad, 1990) links symbols (e.g., words) to the corresponding perceptual information. Hence, robot interactions need to be *grounded*, as meaning depends on the state of the physical world and the interpretation crucially interplays with perception, as pointed out by psycho-linguistic theories (Tanenhaus et al., 1995). The integration of perceptual information derived from the robot's sensors with an ontologically motivated description of the world has been adopted as an augmented representation of the environment, in the so-called *semantic maps* (Nüchter and Hertzberg, 2008). In these maps, the existence of real world objects can be associated to *lexical* information, in the form of entity names given by a knowledge engineer or spoken by a user for a pointed object, as in Human-Augmented Mapping (Diosi et al., 2005; Gemignani et al., 2016). While Command Interpretation for Interactive Robotics has been mostly carried out over the only evidence specific to the linguistic level (see, for example, (Chen and Mooney, 2011; Matuszek et al., 2012)), we argue that a proper Spoken Language Understanding (SLU) for Human-Robot Interaction should be context-aware, in the sense that both the user and the robot live in and make references to a shared environment. For example, in the above command, *"taking"* is the intended action whenever a mug is actually in the kitchen, so that *"the mug in the kitchen"* refers to a single argument. On the contrary, the command may refer to a *"bringing"* action, when no mug is in the kitchen and *the mug* and *in the kitchen*




*Proceedings of the First Workshop on Language Grounding for Robotics*, pages 25–34,
Vancouver, Canada, July 30 - August 4, 2017. ©2017 Association for Computational Linguistics

correspond to different semantic roles. We are interested in an approach for the interpretation of robotic spoken commands that is consistent with (*i*) the world (with all the entities composing it), (*ii*) the Robotic Platform (with its inner representations and capabilities), and (*iii*) the linguistic information derived from the user's utterance.

In this paper, we foster machine learning methodologies for Spoken Language Understanding that force the above research perspective: this is obtained by extending the linguistic evidence that can be extracted from the uttered commands with perceptual evidence directly derived by the semantic map of a robot. In particular, the interpretation process is modeled as a sequence labeling problem where the final labeler is trained by applying Structured Learning methods over realistic commands expressed in domestic environments, as in (Bastianelli et al., 2017). The resulting interpretations adhere to Frame Semantics (Fillmore, 1985): this well-established theory provides a strong linguistic foundations to the overall process while enforcing its applicability, as it is made independent of the vast plethora of existing robotic platforms. Such methodologies have been implemented in a free and ready-to-use framework, here presented, whose name is *LU4R* - an adaptive spoken **L**anguage **U**nderstanding framework for(**4**) **R**obots. LU4R is entirely coded in Java and, thanks to its Client/Server architectural design, it is completely decoupled from the robot, enabling for an easy and fast deployment on every platform[1].

As the aforementioned approaches rely on realistic data, in this work we also present an extended version of *HuRIC* - a **Hu**man **R**obot **I**nteraction Corpus, originally introduced in (Bastianelli et al., 2014) This resource is a collection of realistic spoken commands that users might express towards generic service robots. In this resource, each sentence is labeled with morpho-syntactic information (e.g., dependency relations, POS tags, . . . ), along with its correct interpretation in terms of semantic frames (Baker et al., 1998). In our extension, each annotated sentence is paired with a semantic representation of the world, that justifies the command itself. To the best of our knowledge this is the first corpus providing such a rich representation of a robotic spoken command[2].

This extension of HuRIC supports a broader evaluation of LU4R chain against the information introduced by perceptual knowledge. We observed a significant increase in performance w.r.t. inherent ambiguities of the language, whose outcomes are encouraging for the deployment of such system in realistic applications.

The rest of the paper is structured as follows. Section 2 provides a short survey of existing approaches to SLU for Human-Robot Interaction. Section 3 describes the semantic analysis process that represents the core of LU4R. In Section 4, an architectural description of the entire framework is provided, as well as an overall introduction about its integration with a generic robot. Section 5 describes the extension of HuRIC, while in Section 6 we provide empirical evidence demonstrating the applicability of the proposed system in the interpretation of robotic commands, by reporting our experimental results. In Section 7 we draw some conclusions.

## 2   Related Work

In Robotics, some solutions for the interpretation of spoken commands have been modeled using grammar-based approaches. In general, they provide mechanisms to enrich the syntactic structure with semantic information, to build a semantic representation during the transcription process (Bos, 2002; Bos and Oka, 2007).

Other approaches are based on formal languages, as in (Kruijff et al., 2007; Thomason et al., 2015), where Combinatory Categorial Grammar (CCG) are applied for spoken dialogues in Human-Robot Interaction, and in (Perera and Veloso, 2015) where template-based algorithms allow extracting semantic interpretations of robotic commands by applying specific templates over the corresponding syntactic trees.

Data-driven methods have been also applied to command interpretation for robotic applications. Examples are (MacMahon et al., 2006) and (Chen and Mooney, 2011), where the parsing of route instructions is addressed as a Statistical Machine Translation task between the human language and a synthesized robot language. The same approach is applied in (Matuszek et al., 2010) to learn translation models between natural language and formal descriptions of paths. A probabilistic CCG is used in (Matuszek et al., 2012) to map natu-

---

[1]LU4R        can        be        downloaded        at `http://sag.art.uniroma2.it/lu4r.html`

[2]The extended version of HuRIC will be released at `http://sag.art.uniroma2.it/huric.html`

ral navigational instructions into robot executable commands. The same problem is faced in (Kollar et al., 2010; Duvallet et al., 2013), where Spatial Description Clauses are parsed from sentences through sequence labeling approaches. In (Tellex et al., 2011), the authors address natural language instructions about motion and grasping, that are mapped into Generalized Grounding Graphs ($G^3$). In (Fasola and Mataric, 2013a,b), Spoken Language Understanding (SLU) for pick-and-place instructions is performed through a Bayesian classifier trained over a specific corpus. In (Misra et al., 2016), the authors define a probabilistic approach to ground natural language instructions within a changing environment.

In this paper we present a data-driven approach that integrates an explicit semantic representation with linguistic generalization induced through machine learning. On the one hand, the interpretation is carried out according to the Frame Semantics paradigm (Fillmore, 1985), thus resulting in a principled meaning representation formalism. Moreover, a context-dependent interpretation process is realized: knowledge derived from perceptual evidence is made available and directly used to discriminate against conflicting interpretations. Perceptual information is here represented through an ontologically motivated description of the surrounding environment, i.e., a *semantic map* (Nüchter and Hertzberg, 2008). The semantic map is an explicit representation of the knowledge about surroundings, acquired to enable reasoning over environments, objects and properties. In the map, the existence and position of real world objects is associated to lexical information, in the form of entity class names. On the other hand, machine learning depends on such perceptual information, thus inducing the contextual preconditions of the involved disambiguation choices from real examples, i.e. sentence-map pairs. The process can thus provide different interpretations of one sentence against different maps and realizes a highly reusable and mostly domain-independent model of grounded interpretation.

## 3   The Language Understanding Cascade

A command interpretation system for a robotic platform must produce interpretations of user utterances. In this paper, we consider Frame Semantics (Fillmore, 1985), the formalization promoted in the FrameNet (Baker et al., 1998) project, where actions expressed in user utterances can be modeled as *semantic frames*. Each frame represents a micro-theory about a real world situation, e.g., the actions of *bringing*, *motion* or *manipulation*. Such micro-theories encode all the relevant information needed for their correct interpretation. This information is represented in FrameNet via the so-called *frame elements*, whose role is to specify the participating entities in a frame, e.g., the THEME frame element represents the object that is taken in a *bringing* action.

As an example, let us consider the sentence: "*take the pillow to the couch*". This sentence can be intended as a command whose effect is to instruct a robot that, in order to achieve the task, has to: *(i)* move towards a pillow, *(ii)* pick it up, *(iii)* move to the couch and, finally, *(iv)* release the object on the couch. The language understanding cascade should produce its FrameNet-annotated version:

$$[take]_{Bringing}[the\ pillow]_{\text{THEME}}[to\ the\ couch]_{\text{GOAL}} \quad (1)$$

Semantic frames can thus provide a cognitively sound bridge between the actions expressed in the language and the implementation of such actions in the robot world, namely plans and operations.

The whole SLU process has been designed as a cascade of reusable components, as shown in Figure 1. As we deal with vocal commands, their (possibly multiple) hypothesized transcriptions derived from an Automatic Speech Recognition (ASR) engine constitute the input of this process. It is composed by four modules, whose final output is the interpretation of an utterance, to be used to implement the corresponding robotic actions. First, **Morpho-syntactic analysis** is performed over the available utterance transcriptions by applying morphological analysis, Part-of-Speech tagging and syntactic analysis. In particular, dependency trees are extracted from the sentence as well as POS tags, as shown in Figure 2. Then, if more than one transcription hypothesis is available, the **Re-ranking** module can be activated to compute a new ranking of the hypotheses, in order to get the best transcription out of the initial ranking. This module is realized through a learn-to-rank approach, where a Support Vector Machine exploiting a combination of linguistic kernels is applied, according to (Basili et al., 2013). Third, the best transcription is the input of the **Action Detection** (AD) component. The evoked frames in a sentence are detected, along with the

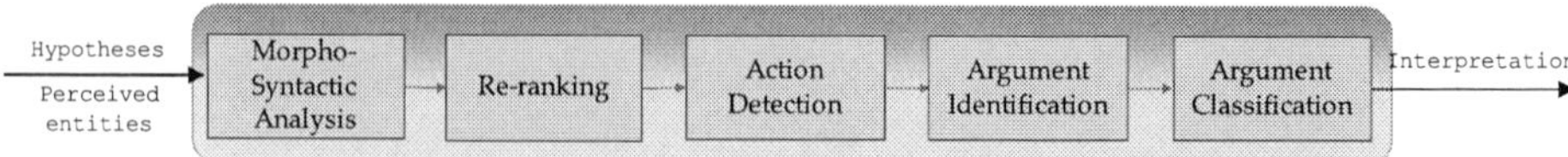


Figure 1: The SLU cascade

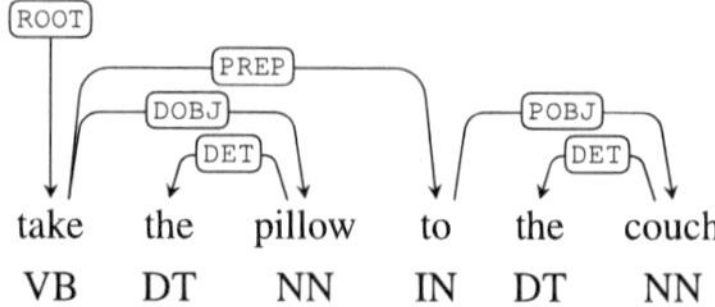


Figure 2: Example of a dependency graph associated to "*take the pillow to the couch*"

corresponding evoking words, the so-called lexical units. Let us consider the recurring sentence: the AD should produce the following interpretation $[take]_{Bringing}$ *the pillow to the couch*. The final step is the **Argument Labeling**, where a set of frame elements is retrieved for each frame. This process is realized in two sub-steps. First, the *Argument Identification* (AI) finds the spans of all the possible frame elements, producing the following form $[take]_{Bringing}$ $[the\ pillow]$ $[to\ the\ couch]$. Then, the *Argument Classification* (AC) assigns the suitable label (i.e., the frame element) to each span thus returning the final tagging shown in the Example (1).

The AD, AI and AC steps are modeled as a sequence labeling task, as in (Bastianelli et al., 2016). The Markovian formulation of a structured SVM proposed in (Altun et al., 2003) is applied to implement the labeler, known as $SVM^{hmm}$. In general, this learning algorithm combines a local discriminative model, which estimates the individual observation probabilities of a sequence, with a global generative approach to retrieve the most likely sequence, i.e., tags that better explain the whole sequence. In other words, given an input sequence $\mathbf{x} = (x_1 \ldots x_l) \in \mathcal{X}$ of feature vectors $x_1 \ldots x_l$, $SVM^{hmm}$ learns a model isomorphic to a $k$-order Hidden Markov Model, to associate $\mathbf{x}$ with a set of labels $\mathbf{y} = (y_1 \ldots y_l) \in \mathcal{Y}$.

A sentence $s$ is here intended as a sequence of words $w_i$, each modeled through a feature vector $x_i$ and associated to a dedicated label $y_i$, specifically designed for each interpretation process[3]: in any case, features combine linguistic evidence

---

[3]More details about the labeling notation can be found in (Bastianelli et al., 2016)

from a targeted sentences, but also properties derived from the semantic map (when available) in order to synthesize information about existence and position of entities around the robot, as discussed in more details in (Bastianelli et al., 2016). During training, the SVM algorithm associates words to step-specific labels: linear kernel functions are applied to different types of features, ranging from linguistic to perception-based features, and linear combinations of kernels are used to integrate independent properties. At classification time, given a sentence $s = (w_1 \ldots w_{|s|})$, the $SVM^{hmm}$ efficiently predicts the tag sequence $\mathbf{y} = (y_1 \ldots y_{|s|})$ using a Viterbi-like decoding algorithm. More details about the construction of feature vectors $x_i$ are reported in (Bastianelli et al., 2016).

Notice that both the re-ranking and the semantic parsing phases can be realized in two different settings, depending on the type of features adopted in the labeling process. It is thus possible to rely upon linguistic information to solve the given task, or also on perceptual knowledge coming from a semantic map. In the first case, that we call *basic* setting, the information used to solve the task comes from linguistic inputs, as the sentence itself or external linguistic resources. These models correspond to the methods discussed in (Bastianelli et al., 2017; Basili et al., 2013). In the second case, the *simple* setting, when perceptual information is made available to the chain, a context-aware interpretation is triggered, as in (Bastianelli et al., 2016). Such perceptual knowledge is mainly exploited through a *linguistic grounding* mechanism. This lexically-driven grounding is estimated through distances between filler (i.e., argument heads) and entity names. Such a semantic distance integrates metrics over word vectors descriptions and phonetic similarity. Word semantic vectors are here acquired through corpus analysis, as in Distributional Lexical Semantic paradigms (Turney and Pantel, 2010). They allow to map referential elements, such as lexical fillers, e.g., *couch*, to entities, e.g., a `sofa`, by thus modeling synonymy or co-hyponymy. Conversely, phonetic similarities are smoothing factors against possible ASR transcription errors, e.g., *pitcher* and *picture*: this allows to actually cope with the noisy phenomena characterizing spoken language.

Once links between fillers and entities have been activated, they act as abductive hypothesis: they inspire features related to individual words that express perceptual information (e.g. presence/absence of referred objects in the environment or spatial relations between them) as well as lexical knowledge (e.g. semantic and phonetic similarity between entity names and uttered references). The labeler trained over such richer descriptions is made thus sensitive to perceptual information both in the learning and the tagging process. As a side effect, the above mechanism provides the robot with the set of linguistically-motivated groundings, that can be potentially used for any further grounding process.

This information can be crucial in the correct interpretation of ambiguous commands, which depends on the specific environmental setting the robot is operating into. A clear example is the command *"bring the pillow on the couch in the living room"*. Such a sentence may have two different interpretations, according to the configuration of the environment. In fact, when the couch is located into the living room, the goal of the *Bringing* action is the couch and interpretation will be: $[bring]_{Bringing}[the\ pillow]_{\text{THEME}}[on\ the\ couch\ in\ the\ living\ room]_{\text{GOAL}}$. Conversely, if the couch is outside the living room, it means that probably the pillow is already on the couch. Hence, the interpretation of the sentence will be different, due to different argument spans, and the couch becomes the goal of the *Bringing* action: $[bring]_{Bringing}[the\ pillow\ on\ the\ couch]_{\text{THEME}}[in\ the\ living\ room]_{\text{GOAL}}$.

Additional details about the pure linguistic approach can be found in (Bastianelli et al., 2017).

## 4  The LU4R Framework

The architecture of the system considers two main actors, as shown in Figure 3: the *Robotic Platform* and *LU4R*, where the processing cascade of the latter component have been introduced in the previous Section.

The Client-Server communication schema between LU4R and the Robot allows for the independence from the Robotic Platform, in order to maximize the re-usability and integration in heterogeneous robotic settings. LU4R exhibits semantic capabilities (e.g., disambiguation, predicate detection or grounding into robotic actions and environments) that are designed to be general enough to be representative of a large set of application scenarios.

It is obvious that an interpretation process must be achieved even when no information about the domain/environment is available, i.e., a scenario involving a *blind* but speaking robot, or when the actions a robot can perform are not made explicit. At the same time, the proposed SLU cascade makes available methods to specialize its semantic interpretation process to individual situations where more information is available about goals, the environment and the robot capabilities. These methods are expected to support the optimization of the core SLU process against a specific interactive robotics setting, in a cost-effective manner. In fact, whenever more information about the environment perceived by the robot (e.g., a semantic map) or about its capabilities is provided, the interpretation of a command can be improved by exploiting a more focused scope.

In order to better understand the different operating modalities of LU4R, some assumptions toward the Robotic Platform must be made explicit: this will allow to precisely establish functionalities and resources that the robot needs to provide to unlock the more complex processes. These information will be used to express the experience that the robot is able to share with the user (i.e., the perceptual knowledge about the environment where the linguistic communication occurs and some lexical information and properties about objects in the environment) and some level of awareness about its own capabilities (e.g., the primitive actions that the robot is able to perform, given its hardware components).

### 4.1  The Robotic Platform

The overall framework contemplates a generic Robotic Platform, whose task, domain and physical setting are not necessarily specified. In order to make the SLU process independent of the above specific aspects, we assume that the platform requires, at least, the following modules: *(i)* an Automatic Speech Recognition (ASR) system, *(ii)* a SLU Orchestrator, *(iii)* a Grounding and Command Execution Engine, and *(iv)* a Physical Robot. The ASR component currently re-

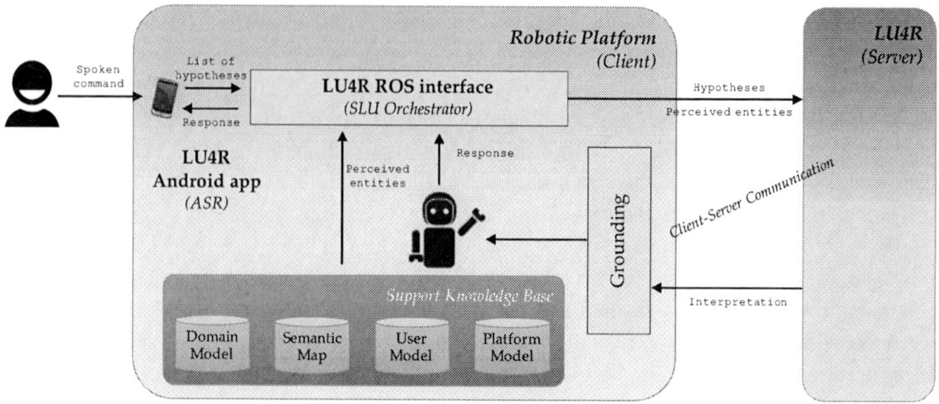

Figure 3: The architecture of the LU4R framework

| | |
|---|---|
| *Number of examples* | 656 |
| *Number of frames* | 18 |
| *Number of predicates* | 767 |
| *Number of roles* | 34 |
| *Predicates per sentence* | 1.17 |
| *Sentences per frame* | 36.44 |
| *Roles per sentence* | 2.04 |
| *Entities per sentence* | 7.29 |

Table 1: Some statistics of the corpus

| | | | |
|---|---|---|---|
| *Motion* | 143 | *Bringing* | 153 |
| *Cotheme* | 39 | *Locating* | 90 |
| *Inspecting* | 29 | *Taking* | 80 |
| *Change_direction* | 11 | *Arriving* | 12 |
| *Giving* | 10 | *Placing* | 52 |
| *Closure* | 19 | *Change_operat_state* | 49 |
| *Being_located* | 38 | *Attaching* | 11 |
| *Releasing* | 9 | *Perception_active* | 6 |
| *Being_in_category* | 11 | *Manipulation* | 5 |

Table 2: Distribution of frames over the corpus

alized exploits the *LU4R Android app* whereas the SLU orchestrator is implemented as a ROS node, through the *LU4R ROS interface*. Additionally, the optional *Support Knowledge Base* component is expected to interface the different involved knowledge sources and support their maintenance: this provides the contextual information discussed above.

## 5 A Perceptual Corpus of Robotic Commands

The computational paradigms adopted here are based on machine learning techniques and depend strictly on the availability of training data. In order to train and test our framework, a proper resource that collects both linguistic and perceptual information is required. To this end, we extended the Human-Robot Interaction Corpus[4] (HuRIC), formerly presented in (Bastianelli et al., 2014), by pairing each English sentence with the corresponding perceptual evidence that justifies the targeted semantics.

HuRIC is based on Frame Semantics and cap-

tures cognitive information about situations and events expressed in sentences. The corpus does not include system or robot-dependent sentences or formalisms. Instead, it contains information strictly related to Natural Language Semantics, decoupled from specific tasks. The corpus exploits different situations representing possible commands given to a robot in a house environment. Each sentence is paired with a set of audio files representing robot commands and its corresponding correct transcription. Each sentence is then annotated with: lemmas, POS tags, dependency trees and Frame Semantics. Semantic frames and frame elements are used to represent the meaning of commands, as they reflect the actions a robot can accomplish in a home environment. In this respect, the AMR representation of the Example 1 is

```
(t1   / take-Bringing
      : Theme (b1 / pillow)
      : Goal (t2 / couch)
)
```

In this way, HuRIC can potentially be used to train all the modules of the processing chain presented in Section 4.

With respect to the previous release, we extended HuRIC by pairing each sentence with the

---

[4]Available at http://sag.art.uniroma2.it/huric. The download page also contains a detailed description of the release format.

corresponding semantic map, composed of all entities populating the environment and presumably "perceived" by the robot. Each entity is represented by the following set of information.

The **Atom** is a unique identifier of the entity, whereas the **Type** of each entity, reflects the class to which each specific entity belongs[5].

The **Preferred Lexical Reference** is used to refer to a class of objects; it is crucial in order to enable the grounding between the commands uttered by the user and the entities within the environment. For example, an entity of the class `table` can be referred by the word *desk*.

Finally, the position of each entity is essential to determine shallow spatial relations between entities, e.g., whether two objects are *near* or *far* from each other. To this end, each entity is associated with its **Coordinate** in the world, in terms of planar coordinates $(x, y)$, elevation $(z)$ and *angle* as the orientation. We adopted a simple numerical scaling that discretized the map.

Table 1 shows the number of annotated sentences, number of frames, along with the average number of entities per sentence. Each entity involved in the command, e.g., *mug* and *kitchen* in the Example 1, is provided with one lexical reference, not necessarily the same word used in the command (e.g. using a synonym such as *cushion* or *sofa*). Detailed statistics about the number of sentences for each frame are reported in Table 2.

## 6 Experimental Evaluation

In order to provide evidence about the benefits of perceptual knowledge, we report an evaluation of the interpretation process of robotic commands over the enhanced version of HuRIC, i.e., contemplating the semantic maps for each sentence.

Table 3 shows the results obtained. The results, expressed in terms of *Precision*, *Recall* and *F1 measure*, focus on the semantic interpretation process, in particular Action Detection (AD), Argument Identification (AI) and Argument Classification (AC) steps, addressing two possible configurations: a basic setting where only linguistic information is exploited (i.e., *noSM*, as the semantic maps are ignored), and the configuration where semantic maps are included into the learning loop (i.e., *SM*). F1 scores measure the quality of a specific module. While in the AD step the F1 refers

---

[5]Notice that an entity can be an object, e.g., *couch*, *pillow*, or a location, e.g., *bedroom*

| | Precision | Recall | F1-Measure |
|---|---|---|---|
| | **AD** | | |
| *noSM* | $94.73 \pm 1.21$ | $94.02 \pm 1.51$ | $94.37 \pm 1.00$ |
| *SM* | $95.69 \pm 1.40$ | $96.90 \pm 1.90$ | **$96.29 \pm 1.56$** |
| | **AI** | | |
| *noSM* | $88.95 \pm 2.24$ | $88.22 \pm 2.08$ | $88.57 \pm 1.65$ |
| *SM* | $91.34 \pm 1.73$ | $91.72 \pm 1.14$ | **$91.53 \pm 1.43$** |
| | **AC** | | |
| *noSM* | $93.05 \pm 1.05$ | $93.05 \pm 1.05$ | $93.05 \pm 1.05$ |
| *SM* | $94.02 \pm 1.25$ | $94.02 \pm 1.25$ | **$94.02 \pm 1.25$** |

Table 3: Experimental evaluation of the semantic interpretation process

to the ability to extract the correct frame(s) (i.e., robot action(s) expressed by the user) evoked by a sentence, in the AI step it evaluates to the correctness of the predicted argument spans. Finally, in the AC step the F1 measures the accuracy of the classification of individual arguments. The experiments have been performed in a 5-fold cross validation setting. In this respect, Table 3 provides also the *standard deviations* among the different folds. We tested each sub-module in isolation, feeding each step with gold information provided by the previous step in the chain. Moreover, the evaluation has been carried out considering the correct transcriptions, i.e., not contemplating the error introduced by the Automatic Speech Recognition system.

The overall results are encouraging for the application of the proposed approach in realistic scenarios. In fact, the F1 is always higher than 94% in the recognition of semantic predicates used to express intended actions (AD). The system is able to recognize the involved entities (AC) with high accuracy as well, with a F1 higher than 93% in both *noSM* and *SM* settings. This result is surprising when analyzing the complexity of the task. In fact, the classifier is able to cope with a high level of uncertainty, as the amount of possible semantic roles is sizable, i.e., 34. In general, the most challenging task seems to be the ability to recognize the spans composing a single frame element (AI).

Regarding the *noSM* setting, i.e., only linguistic information, one of the most frequent error concerns the ambiguity of the *"take"* verb. In fact, as explained in the previous sections, the interpretation of such verb may be different (i.e., either *Bringing* or *Taking*), depending on the configuration of the environment. As this particular setting does not provide any kind of perceptual information, the system is not able to correctly discriminate among them. Hence, the resulting interpretation will be wrong, as it does not reflect the semantics that is motivated by the environment. In terms of F1 measure, this issue affects mainly the Argument Identification step (AI), rather than the Action Detection (AD) one, as for each (possibly) wrong frame, there could be more than two (possibly) wrong arguments. For example, the sentence *"take the mug in the kitchen"* will be probably recognized to be a *Taking* action, even though it is labeled as *Bringing*, i.e., *mug* and *kitchen* are supposed to be far in the environment. While the AD step will receive just one penalty for the wrong recognized action, the AI step is penalized twice, as two arguments were expected by the gold standard annotation, i.e., the *the mug* as THEME and the *in the kitchen* as GOAL, instead of one, i.e., *the mug in the kitchen* as a single THEME argument.

When looking at the *SM* setting, it seems that the injection of perceptual knowledge into the semantic analysis process is able to mitigate the effect of the aforementioned phenomena and each SLU step gains in predictive performance. In the case of AD, the information about the entities shows a relative improvement of $+2.03\%$ in terms of F1 ($94.37\%$ vs $96.29\%$). This means that the semantic map allows to predict the intended action more accurately, whenever the underlying semantic ambiguity depends on the configuration of the environment. The tight correlation between the predicted action and the frame elements suggests a similar behavior in Argument Identification. In fact, as well as for the AD, in the AI step perceptual knowledge reveal its support in predicting the correct spans of semantic arguments, with a relative improvement of $+3.34\%$ w.r.t. the F1 score. Though a lower gain is observed ($+1.04\%$), the introduction of Distributional Semantics improves the ability of recognizing the correct frame element for a given argument span, i.e., AC step. This is probably due to the lexical generalization provided by the word embeddings, whenever alternative naming are used to refer to an entity of the semantic map.

Finally, small values of standard deviation suggest that the system seems to be rather stable across the different iterations of the experiment and that the results do not depend on specific splits of the entire dataset.

## 7  Conclusions

In this paper, we presented a comprehensive framework for the design of robust natural language interfaces for Human-Robot Interaction (HRI). The corresponding implementation is specifically designed for the automatic interpretation of spoken commands in domestic environments. The proposed solution relies on Frame Semantics and supports a structured learning approach to language processing able to map individual sentence transcriptions to meaningful commands. A hybrid discriminative and generative learning method is proposed to map the interpretation process into a cascade of sentence annotation tasks. The interpretation of commands is made dependent on the robot's environment; in fact the adopted training annotations not only express linguistic evidence from source utterances, but also account for specific perceptual information derived from a reference map. In this way the semantic map aspects useful to interpretation are expressed via feature modeling with the structured learning mechanism applied. Such perceptual knowledge is derived from a semantically-enriched implementation of a robot map, i.e., its semantic map. It expresses information about the existence and position of entities surrounding the robot: as this is also available to the user, this information is crucial to disambiguate predicates and role assignments.

To this end, we trained the machine learning processes by using an extended version of *HuRIC*, the Human Robot Interaction Corpus. This corpus, originally composed by sentences in English, now benefits from the introduction of such semantic maps, expressed as lists of entities and supporting the research in natural language interfaces for Robots in such language. The empirical results obtained over the perceptual version of the dataset show a significant improvement w.r.t. the pure linguistic process. This confirms the effectiveness of the proposed processing chain.

Future research will also focus on the extension of the proposed methodology, e.g., by considering spatial relations between entities in the environment or their physical characteristics, such as their color and the application of this solution in interactive question answering or dialogue with robots.

**References**

Yasemin Altun, I. Tsochantaridis, and T. Hofmann. 2003. Hidden Markov support vector machines. In *Proc. of ICML*.

Collin F. Baker, Charles J. Fillmore, and John B. Lowe. 1998. The berkeley framenet project. In *Proceedings of ACL and COLING*. pages 86–90.

Roberto Basili, Emanuele Bastianelli, Giuseppe Castellucci, Daniele Nardi, and Vittorio Perera. 2013. Kernel-based discriminative re-ranking for spoken command understanding in hri. In *AI* IA 2013: Advances in Artificial Intelligence*, Springer International, pages 169–180.

Emanuele Bastianelli, Giuseppe Castellucci, Danilo Croce, Roberto Basili, and Daniele Nardi. 2014. Huric: a human robot interaction corpus. In *Proc. of LREC 2014*. Reykjavik, Iceland.

Emanuele Bastianelli, Giuseppe Castellucci, Danilo Croce, Roberto Basili, and Daniele Nardi. 2017. Structured learning for spoken language understanding in human-robot interaction. *The International Journal of Robotics Research* 0(0):To appear in. https://doi.org/10.1177/0278364917691112.

Emanuele Bastianelli, Danilo Croce, Andrea Vanzo, Roberto Basili, and Daniele Nardi. 2016. A discriminative approach to grounded spoken language understanding in interactive robotics. In *Proc. of the 25th IJCAI, New York*.

Johan Bos. 2002. Compilation of unification grammars with compositional semantics to speech recognition packages. In *Proceedings of the 19th International Conference on Computational Linguistics - Volume 1*. Association for Computational Linguistics, Stroudsburg, PA, USA, COLING '02, pages 1–7. https://doi.org/10.3115/1072228.1072323.

Johan Bos and Tetsushi Oka. 2007. A spoken language interface with a mobile robot. *Artificial Life and Robotics* 11(1):42–47.

David L. Chen and Raymond J. Mooney. 2011. Learning to interpret natural language navigation instructions from observations. In *Proc. of the 25th AAAI Conference*. pages 859–865.

Albert Diosi, Geoffrey R. Taylor, and Lindsay Kleeman. 2005. Interactive SLAM using laser and advanced sonar. In *Proc. of the 2005 International Conference on Robotics and Automation*. pages 1103–1108.

Felix Duvallet, Thomas Kollar, and Anthony Stentz. 2013. Imitation learning for natural language direction following through unknown environments. In *2013 IEEE International Conference on Robotics and Automation, Karlsruhe, Germany, May 6-10, 2013*. pages 1047–1053. https://doi.org/10.1109/ICRA.2013.6630702.

Juan Fasola and Maja J Mataric. 2013a. Using semantic fields to model dynamic spatial relations in a robot architecture for natural language instruction of service robots. In *International Conference on Intelligent Robots and Systems (IROS), 2013 IEEE/RSJ*. pages 143–150.

Juan Fasola and Maja J. Mataric. 2013b. Using spatial semantic and pragmatic fields to interpret natural language pick-and-place instructions for a mobile service robot. In *Social Robotics: 5th International Conference, ICSR 2013, Bristol, UK, October 27-29, 2013, Proceedings*, Springer International Publishing, pages 501–510.

Charles J. Fillmore. 1985. Frames and the semantics of understanding. *Quaderni di Semantica* 6(2):222–254.

Guglielmo Gemignani, Roberto Capobianco, Emanuele Bastianelli, Domenico Daniele Bloisi, Luca Iocchi, and Daniele Nardi. 2016. Living with robots. *Robot. Auton. Syst.* 78(C):1–16. https://doi.org/10.1016/j.robot.2015.11.001.

S. Harnad. 1990. The symbol grounding problem. *Physica D: Nonlinear Phenomena* 42(1-3):335–346.

Thomas Kollar, Stefanie Tellex, Deb Roy, and Nicholas Roy. 2010. Toward understanding natural language directions. In *Proceedings of the 5th ACM/IEEE*. Piscataway, NJ, USA, HRI '10, pages 259–266.

Geert-Jan M. Kruijff, H. Zender, P. Jensfelt, and Henrik I. Christensen. 2007. Situated dialogue and spatial organization: What, where... and why? *International Journal of Advanced Robotic Systems* 4(2).

Matt MacMahon, Brian Stankiewicz, and Benjamin Kuipers. 2006. Walk the talk: connecting language, knowledge, and action in route instructions. In *proceedings of the 21st national conference on Artificial intelligence - Volume 2*. AAAI Press, AAAI'06, pages 1475–1482.

Cynthia Matuszek, Dieter Fox, and Karl Koscher. 2010. Following directions using statistical machine translation. In *Proceedings of the 5th ACM/IEEE International Conference on Human-robot Interaction*. IEEE Press, Piscataway, NJ, USA, HRI '10, pages 251–258.

Cynthia Matuszek, Evan Herbst, Luke S. Zettlemoyer, and Dieter Fox. 2012. Learning to parse natural language commands to a robot control system. In Jaydev P. Desai, Gregory Dudek, Oussama Khatib, and Vijay Kumar, editors, *ISER*. Springer, volume 88 of *Springer Tracts in Advanced Robotics*, pages 403–415.

Dipendra K. Misra, Jaeyong Sung, Kevin Lee, and Ashutosh Saxena. 2016. Tell me dave: Context-sensitive grounding of natural language to manipulation instructions. *The International Journal of Robotics Research* 35(1-3):281–300. https://doi.org/10.1177/0278364915602060.

Andreas Nüchter and Joachim Hertzberg. 2008. Towards semantic maps for mobile robots. *Robot. Auton. Syst.* 56(11):915–926.

Vittorio Perera and Manuela M. Veloso. 2015. Handling complex commands as service robot task requests. In *Proceedings of the Twenty-Fourth International Joint Conference on Artificial Intelligence, IJCAI 2015, Buenos Aires, Argentina, July 25-31, 2015*. pages 1177–1183.

M. Tanenhaus, M. Spivey-Knowlton, K. Eberhard, and J. Sedivy. 1995. Integration of visual and linguistic information during spoken language comprehension. *Science* 268:1632–1634.

Stefanie Tellex, Thomas Kollar, Steven Dickerson, Matthew R Walter, Ashis Gopal Banerjee, Seth Teller, and Nicholas Roy. 2011. Approaching the symbol grounding problem with probabilistic graphical models. *AI Magazine* 34(4):64–76.

Jesse Thomason, Shiqi Zhang, Raymond Mooney, and Peter Stone. 2015. Learning to interpret natural language commands through human-robot dialog. In *Proceedings of the 24th International Conference on Artificial Intelligence*. AAAI Press, IJCAI'15, pages 1923–1929. http://dl.acm.org/citation.cfm?id=2832415.2832516.

Peter D. Turney and Patrick Pantel. 2010. From frequency to meaning: Vector space models of semantics. *J. Artif. Int. Res.* 37(1):141–188. http://dl.acm.org/citation.cfm?id=1861751.1861756.

# Natural Language Grounding and Grammar Induction
## for Robotic Manipulation Commands


**M Alomari, P Duckworth, M Hawasly, D C Hogg, A G Cohn**
{scmara, scpd, m.hawasly, d.c.hogg, a.g.cohn}@leeds.ac.uk,
School of Computing, University of Leeds, Leeds LS2 9JT, England



## Abstract

We present a cognitively plausible system capable of acquiring knowledge in language and vision from pairs of short video clips and linguistic descriptions. The aim of this work is to teach a robot manipulator how to execute natural language commands by demonstration. This is achieved by first learning a set of visual 'concepts' that abstract the visual feature spaces into concepts that have human-level meaning. Second, learning the mapping/grounding between words and the extracted visual concepts. Third, inducing grammar rules via a semantic representation known as Robot Control Language (RCL). We evaluate our approach against state-of-the-art supervised and unsupervised grounding and grammar induction systems, and show that a robot can learn to execute never seen-before commands from pairs of unlabelled linguistic and visual inputs.


## 1 Introduction

Understanding natural language commands is essential for robotic systems to naturally and effectively interact with humans. In this paper, we present a framework for learning the linguistic and visual components needed to enable a robot manipulator of executing new natural language commands in a table-top environment. The learning is divided into three steps: ($i$) learning of visual concepts, ($ii$) mapping the words to the extracted visual concepts (i.e. language grounding), and ($iii$) inducing grammar rules to model the natural language sentences. Our system updates its knowledge in language and vision incrementally, by processing a pair of inputs at a time. The input to our system consists of a short video clip of a

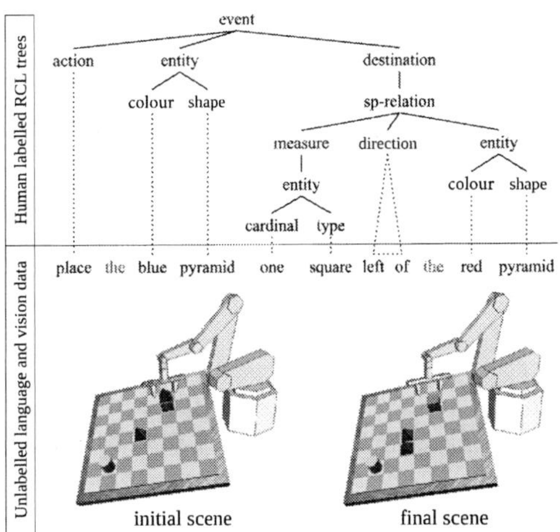


Figure 1: Human expert annotation of a natural language command. The annotation includes grounding for each word and an RCL tree.

robot performing a single action, e.g. a *pick up* or a *move* action, paired with a natural language command corresponding to the action in the video. The natural language commands were collected from volunteers and online crowd-sourcing tools such as Amazon Mechanical Turk with minimal amount of supervision or constraints on the language structure which annotators could use.

Generally, *supervised* language grounding and grammar induction systems learn from sentences that have been *manually* annotated by a human expert. As shown in Fig. 1, each word gets annotated with a semantic category (e.g. colour, shape, etc.), and the grammar structure gets annotated using a tree that connects the different words together (e.g. RCL trees) as presented by Dukes (2014) and Matuszek (2013). The manual annotation of data is a labour intensive task that hinders learning from large corpora, and such labels are not necessarily available for all languages. Therefore, *unsupervised* grounding and grammar induction systems




*Proceedings of the First Workshop on Language Grounding for Robotics*, pages 35–43,
Vancouver, Canada, July 30 - August 4, 2017. ©2017 Association for Computational Linguistics

learn language models from unlabelled/raw linguistic data by exploiting co-occurrences of words in a corpus, which generally performs poorly. Therefore, in this work we take a different approach to learn words meanings and grammar rules by connecting natural language to extracted visual features from video clips.

## 2 Related Work

**Language acquisition** has been a long standing objective of AI and cognition research. Siskind (1996) was one of the earliest researchers to try to understand in a computational setting how children learn their native language and map it to their vision. Following his research, in the field of developmental robotics, researchers have connected language and vision to teach their robots different concepts; one of the earliest works in the field is a system by Roy *et al.* (1999) where a robot capable of learning audio-visual associations (e.g. objects' names) using mutual information criterion was presented. Several robotic applications were developed subsequently, such as Steels (2001) where language games for autonomous robots are used to teach the meaning of words in a simple static world. Further, researchers developed systems capable of learning objects' names and spatial relations by interacting with a human or robot teacher, as by Steels (2002), Bleys (2009) and Spranger (2015). Providing machines with the ability to understand natural language commands is a key component for a natural human-robot interaction. For example, "Back to the blocks world" (She et al., 2014) and "Tell me Dave" (Misra et al., 2015) focused on learning the natural language commands for simple manipulation tasks. This is similar to our work, but we improve on their work in three different aspects. First, their works use a pre-trained language parser to extract relevant words from sentences for learning, while we learn from raw (unprocessed) linguistic inputs. Second, they assume the robot knows the visual representations of shapes, spatial-relations and actions beforehand, while we learn these automatically from videos. Finally, we learn the grammar rules along with word groundings.

**Language grounding** systems in robotic applications are usually trained in a supervised setting on a corpus of labelled/tagged text as in Tellex *et al.* (2011), Bollegal *et al.* (2015), and Cui *et al.* (2016). The manual annotation of text is a labour intensive task. Therefore, researchers developed unsupervised techniques that learn the semantic categories of words from unlabelled data by exploiting regularities in natural language as in Schütze (1998), Biemann (2009), Socher *et al.* (2012), and Houthooft *et al.* (2016). Similarly, in **grammar induction**, parsers are commonly trained in a supervised setting on a corpus of annotated grammar trees as presented by Matuszek *et al.* (2013), and Dukes (2014). Other researchers have tackled unsupervised grammar induction from unlabelled sentences as presented by Klein *et al.* (2002), Smith and Eisner (2005), Barzilay *et al.* (2009), Chen *et al.* (2011), Ponvert *et al.* (2011), and Søgaard (2012). While unsupervised grounding and grammar induction techniques enable learning from unlabelled data, their performance is usually significantly worse than those of the supervised techniques. In this work, we present a novel technique capable of acquiring grounding and grammar knowledge comparable to supervised techniques from unlabelled data by mapping words to automatically extracted visual concepts from video clips.

## 3 Learning Visual Concepts ($\mathcal{C}$)

In this section, we describe how we represent the visual input data: we first extract a set of visual features from each video clip; then, we show how we abstract values from these features to form a set of clusters (or visual concepts). These clusters are used to learn the visual representation of words in the language grounding section.

We start by processing the video clips to detect and track the objects in each frame. The objects are detected using a table-top object detector (Muja and Ciocarlie, 2013), where each object in a video is assigned a unique *id* (a number), and its location is tracked using a particle filter (Klank et al., 2009). Next, we obtain three sets of observations from each video clip; ($i$) object features: $\{colour, shape, location\}$ of each object, ($ii$) relational features: $\{direction, distance\}$ of each pair of objects in the scene, and ($iii$) the $\{atomic\ actions\}$ that the robot applies on each object during the video. The features and atomic actions are presented in Fig. 2. These features and actions are obtained at every frame in each video. It is worth noting that these features are not intended to be exhaustive, but rather to demonstrate our approach; more features can be added as an extension without changing the learning framework.

**Object**

- $F_{colour}$ : $object \rightarrow ([0, 360) \times [0, 1] \times [0, 1])$; $F_{colour}(o)$ gives HSV colour values for an object.
- $F_{shape}$ : $object \rightarrow \mathbb{R}^{32}$; $F_{shape}(o)$ gives a 32 bin fast point feature histogram (FPFH) per object.
- $F_{location}$ : $object \rightarrow \mathbb{R} \times \mathbb{R} \times \mathbb{R}$; $F_{location}(o)$ gives an $x, y, z$ location wrt the bottom left corner of the table.

**Relational**

- $F_{distance}$ : $object \times object \rightarrow \mathbb{R}$; $F_{distance}(o_1, o_2)$ gives the distance between $o_1$ and $o_2$ centroids.
- $F_{direction}$ : $object \times object \rightarrow [0, 360) \times [0, 360)$; $F_{direction}(o_1, o_2)$ gives the azimuth and altitude angles from $o_1$ to $o_2$.

**Atomic Actions**

- $F_{action}$ : $object \rightarrow [approach, depart, grasp, discard, lift, lower, move, rotate, null]$; $F_{action}(o)$ gives the atomic action, as shown below.

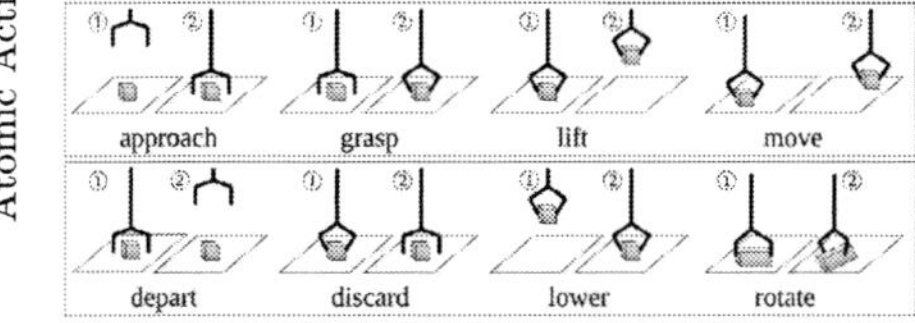


Figure 2: Predefined features and atomic actions.

Once the observations (objects, relations, actions) have been obtained for all objects in all video clips, we process them to extract the unique concepts, e.g. distinguishing the shape *cube* from the shape *prism*, and the action *pick up* from *put down*, etc. This is achieved by clustering the values of each feature space separately to obtain multiple clusters in the dataset. The extracted clusters are used to construct a visual concept vector ($\mathcal{C}$) with length equal to the total number of clusters. This forms the list of possible visual representations of words. For instance, Dukes (2013) dataset (intended to train semantic taggers in a supervised setting), contains four unique shapes: prism, cube, ball, and cylinder. We cluster the shape values of all objects from all video clips and are thus able to extract these clusters/shapes, e.g. $shape_1$ = cube, $shape_2$ = prism, $shape_3$ = ball, and $shape_4$ = cylinder. The clustering is performed using a combination of Gaussian Mixture Models and Bayesian Information Criterion to find the optimal number of clusters representing the data in each feature space. The same clustering method is used on all observations (*colours, locations, directions, distances, and atomic actions*) each of which is done separately, then the outputs (or the clusters) are combined into a single vector $\mathcal{C} = \{shape_1, shape_2, shape_3, shape_4, colour_1, \ldots, location_1, \ldots, direction_1, \ldots, distance_1, \ldots, action_1, \ldots\}$ to give the visual representations of words. This is used in the next section for language grounding. After generating the vector $\mathcal{C}$, we go through

each of the video clips and represent the observed contents of each clip (objects, relations, actions) as a collection of entries or predicates. For example, an object with $id = 3$ and shape $shape_1$ ('cube') is represented as the entry $shape_1(3)$. An example from the dataset is shown in Fig. 3 where the entries are shown on the right.

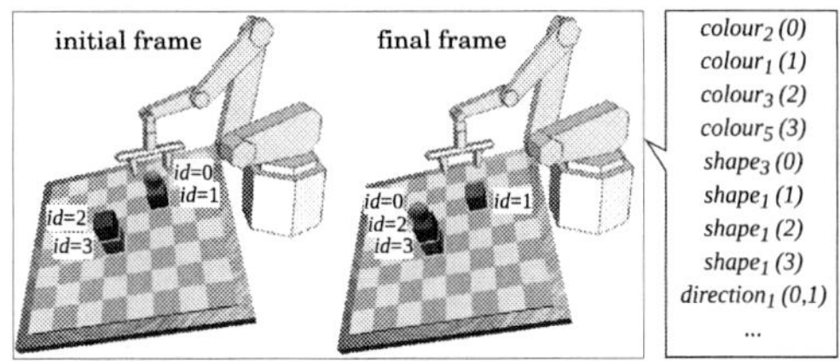


Figure 3: The initial and final frames of a scene from the Dukes (2013) dataset represented in predicates using the learned clusters in $\mathcal{C}$.

## 4  Language Grounding ($\Phi$)

Assigning a word to its correct visual representation is an essential preprocessing step for understanding and executing natural language commands in robotics. In this section, we show how we connect words to their visual representations that we extracted in the previous section. The problem statement of this section is: given (1) a corpus of $n$ sentences $S = \{s_1, \ldots, s_n\}$ that contains $m$ unique words $W = \{w_1, \ldots, w_m\}$, and (2) corresponding video clips $V = \{v_1, \ldots, v_n\}$ that contain $k$ extracted visual concepts $\mathcal{C} = \{c_1, \ldots, c_k\}$, find a partial function $\Phi$ that maps words from language to their representations in vision, $\Phi : W \rightarrow \mathcal{C}$. This language grounding learning problem is formulated as an assignment problem where words $w_i \in W$ should be assigned to clusters $c_j \in \mathcal{C}$ subject to a cost function $\mathcal{F} : W \times \mathcal{C}$ that needs to be minimised. We define the cost function as $\mathcal{F}_{w,c} = (1 - (N_{w,c}/N_w))$, where $N_{w,c}$ is the total number of times a word $w$ and a cluster $c$ appear together, and $N_w$ is the total number of times the word $w$ appears in the entire dataset. This cost function is equal to zero if $w$ and $c$ always appear together, and equal to one if they are never seen together. This provides a clear indication of whether a word $w$ should be mapped to a cluster $c$ or not.

Once the cost function is computed for all word-cluster pairs, we create a cost matrix with words $W$ as rows and clusters $\mathcal{C}$ as columns, as shown in Fig. 4 (left). We then use the Hungarian algorithm

(Kuhn, 1955) to find the grounding for each word by assigning it to its most likely visual concept, i.e. a word can only have one meaning (though multiple words could have the same meaning), as shown in Fig. 4 (right).

To simplify the learning of language grounding in NLP applications, it is common to use a stop word list to remove function words, such as 'the' and 'as', from all sentences. However, since we learn from unlabelled data (i.e. avoiding human annotation including stop word lists), we remove such words by setting a threshold on the Hungarian algorithm. This has the same effect as using term frequency-inverse document frequency (tf-idf) weighting to remove function words (Jones 1972).

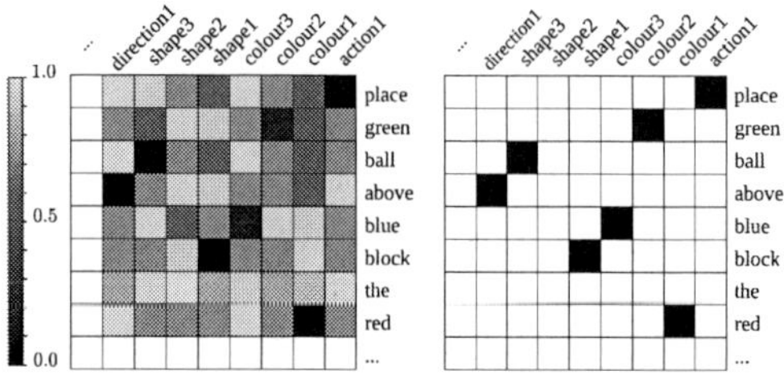

Figure 4: (left:) The cost matrix. (right:) The output of the Hungarian algorithm.

## 5  Generation of RCL Trees ($\Omega$)

Robot Control Language (RCL) is a tree semantic representation for natural language. As shown in Fig. 1, a sentence is represented as an RCL tree where leaf nodes align to the words in the sentence, and non-leaves are tagged using a predefined set of categories that a robot can understand/execute as presented by Matuszek (2013) and Dukes (2013). Although the RCL used in this work is designed to operate within the context of robot manipulation only, it can be extended to other domains such as robot navigation, learning from YouTube how-to videos (Alayrac et al., 2016), or learning cooking instructions (Malmaud et al., 2015). Table 1 lists the different types of RCL elements that are used to compose natural language commands.

In the literature, the problem of transforming sentences into RCL trees has been formulated as a grammar induction one. A parser is trained on pairs of sentences and corresponding human annotated RCL trees, as shown in Fig. 1. The parser is then used to parse new (previously unseen) sentences into RCL trees. The human annotation of RCL trees is labour-intensive, and it would prevent the

| RCL element | Description |
| --- | --- |
| event | Specification of a single command. Takes (action, entity, destination) as children. |
| action | Aligned to a verbal group in NL, e.g. 'place'. |
| entity | Specification of a single entity. Takes (colour, shape, location, sp-relation) as children. |
| sp-relation | Used to specify a spatial relation between two entities or to describe a location. Takes (direction, distance, entity) elements as children. |
| destination | A spatial destination. Takes (sp-relation, location) as children. |
| colour | Colour attribute of an entity, e.g. 'red'. |
| shape | Shape attribute of an entity, e.g. 'pyramid'. |
| location | Location attribute of an entity, e.g. 'centre'. |
| direction | Direction relation between two entities. |
| distance | Distance relation between two entities. |

Table 1: Universal semantic elements in RCL.

robot from learning without constant supervision.

In this paper, we automatically generate a *vision tree* $\Omega_i$ from each video clip $v_i \in V$. We define a *vision tree* as an event tree with only three elements (*action*, *entity*, *destination*), as shown in the example in Fig. 5. The *action*-element holds the action feature extracted from the video $v_i$, the *entity*-element has the *id* of the object that is manipulated by the robot in the video, and the *destination*-element has the final $(x, y, z)$ location feature of that object. In the next section, we show how to use the vision trees $\Omega = \{\Omega_1, \ldots, \Omega_n\}$ to automatically generate language RCL trees analogous to the one in Fig. 1.

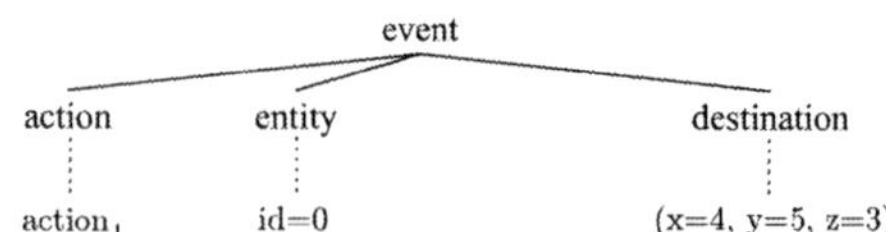

Figure 5: An example of a *vision tree* from the video clip shown in Fig. 3.

## 6  Grammar Induction ($G$)

Grammar induction refers to the process of learning a formal grammar (usually as a collection of re-write rules or productions) from a set of observations. In this work, we show how we learn such rules by mapping natural language commands to visual features. The main contribution of our grammar induction approach is that we *automatically* generate training examples similar to those annotated by a human expert shown in Fig. 1. This is achieved by exploiting the learned groundings $\Phi$ (shown in Fig 4) and the extracted vision trees $\Omega$ (shown in Fig. 5) to successfully replace the human annotator. We formulate the automatic generation of language RCL trees into a search problem as follows. Given (1) vision trees $\Omega = \{\Omega_1, \ldots, \Omega_n\}$, (2) the learned grounding $\Phi : W \to C$, and (3) input sentences $S = \{s_1, \ldots, s_n\}$, we want to search the space of all possible language RCL trees from a sentence $s_i \in S$ for one that matches the extracted vision tree $\Omega_i \in \Omega$. Given a match, we use that language tree to learn grammar $G$. We say a language RCL tree matches a vision tree if the values of all corresponding elements are equal. The procedure to perform the search is divided into five steps (*substitute, connect, query, match,* and *learn*) shown in Algorithm 1. The following sections walk through an example of the entire search process.

---

**Algorithm 1** Automatic generation of language RCL trees

---

1: **procedure** SEARCH ALGORITHM
2: **Inputs** $\Phi, \Omega, S$
3: **Output** $G$
4: **for** each sentence $s_i \in S$ **do**
5:     Substitute each word in $s_i$ with its visual concepts in $\Phi$
6:     Connect vision concepts to create language RCL elements.
7:     Query the RCL elements with video $v_i$.
8:     Match RCL elements with the vision RCL tree $\Omega_i$.
9:     **if** all RCL elements match with $\Omega_i$ **then**
10:         Use RCL elements to learn grammar $G$

---

### 6.1 Substituting words with visual concepts

For each sentence $s_i \in S$ consisting of $t$ words $s_i = \{w_1, \ldots, w_t\}$, we substitute each word with its visual concept using the mapping function $\Phi$ learned in the language grounding section. For instance, a sentence $s_i = \langle place, the, green, ball, above, the, blue, block \rangle$, is transformed using the mapping function $\Phi$ into $s_i' = \langle action_1, None, colour_2, shape_3, direction_1, None, colour_3, shape_1 \rangle$, as shown in Fig. 6 (*substitute*).

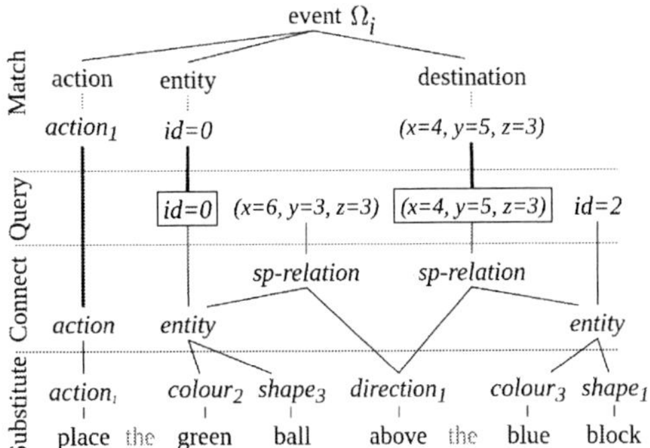


Figure 6: Example for the generation of language RCL tree.

### 6.2 Connecting concepts to generate RCL elements

We group the visual concepts in $s_i'$ to create all possible *entity, action,* or *sp-relation* RCL elements (specified in Table. 1). Particularly, consecutive *colour, shape,* and *location* concepts are grouped to form *entity* RCL elements, consecutive *action* concepts form *action* RCL elements, and consecutive *direction* and *distance* concepts with *entity* elements are grouped to form *sp-relation* RCL elements. For example, in the sentence $s_i' = \langle action_1, colour_2, shape_3, direction_1, colour_3, shape_1 \rangle$ the visual concept $colour_2$ and $shape_3$ are grouped together to generate an *entity* element $entity(colour_2, shape_3)$. The same procedure applies to *action* and *sp-relation* elements, as shown in Fig. 6 (*Connect*).

### 6.3 Querying RCL elements

For each connected *entity* and *sp-relation* RCL element from the previous step, we query the observations in the scene to retrieve a corresponding object $id$ or an $(x, y, z)$ location. For instance, the element $entity(colour_2, shape_3)$ matches from the extracted observations shown in Fig. 3 (right) a single object with $id = 0$, as it is the only object that satisfies both query properties $colour_2$ and $shape_3$ ('green ball' in this case). Similarly, querying the $sp-relation(direction_1, entity(colour_3, shape_1))$ returns $(x = 4, y = 5, z = 3)$ referring to 'above blue block'. This is repeated for all connected entities and spatial relations, as shown in Fig. 6 (*Query*).

If multiple objects in the scene satisfy a query, a list of *ids* is returned, while if there are none the query returns an empty list (this might happen due to noise in vision and/or language). In the results section we show that our system is capable of learning using these connections, even in the presence of noise from real-world data.

### 6.4 Matching RCL elements with vision RCL trees $\Omega$

Given the results of the previous process, we match the returned query results to elements from the vision RCL trees in $\Omega$. For example, the vision RCL tree $\Omega_i$ in Fig. 5 has an *entity* element with $id = 0$, matching the query output in Fig. 6, thus matching it with 'green ball'. This is repeated for *action* and *destination* elements in $\Omega_i$. This process grounds linguistic descriptions to their visual counterparts in RCL without human supervision as shown in Fig. 6 (*Match*).

### 6.5 Learning grammar G

To provide a robot with the ability of understanding natural language commands, we learn grammar $G$ from the automatically generated language RCL trees. The grammar induction is performed using probabilistic context free grammar (Charniak, 1997), by training a semantic parser on the automatically generated examples. The parser is then used in the experiment section to parse new natural language commands into Robot Control Language (RCL) trees. This concludes the search process.

## 7 Experimental Procedure

We evaluate the performance of our system using two datasets; a synthetic-world dataset, and a new, simplified real-world dataset of table-top environment.

For the **synthetic-world**, we use the *Train Robots* dataset[1] which was designed to develop systems capable of understanding verbal spatial commands described in a natural way (Dukes, 2013). Non-expert users from Amazon Mechanical Turk were asked to annotate appropriate natural language commands to 1000 different scenes. A total of 4850 commands were collected and later annotated by human experts with appropriate RCL trees. Examples from this synthetic dataset are shown in Fig. 1 and 3.

For the **real-world** setup, we use a Baxter robot as our test platform and attach a Microsoft Kinect2 sensor to its chest as shown in Fig. 7. The Kinect2

device is used to collect RGBD videos as volunteers controlled the robot arm to perform various manipulation tasks with real objects from the robot's point of view. The dataset consists of 204 videos with $17,373$ frames in total. The videos are annotated with 1024 natural language commands (5 per video in average) by a separate group of volunteers[2]. A total of 51 different objects are manipulated in the videos such as basic block shapes, fruit, cutlery, and office supplies. A detailed description of both datasets is presented in Table. 2.

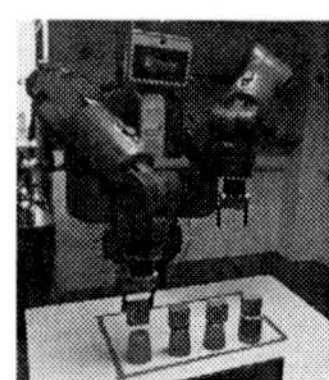

Figure 7: Example scenes from our robotic dataset.

| Dataset contents | | | | | | | |
|---|---|---|---|---|---|---|---|
| features | $A$ | $B$ | $C$ | $D$ | $E$ | $F$ | $G$ |
| Synthetic | 9 | 4 | 4 | 3 | $NA$ | 4 | 24.8 |
| Real-world | 11 | 13 | 3 | 5 | 2 | 3 | 5.3 |

Table 2: Number of concepts in A-colour, B-shape, C-location, D-direction, E-distance, and F-action features in both datasets, and G-average number of objects present in each scene.

### 7.1 Implementation Details

For the real-world dataset, **objects** are detected using a tabletop object detector on the first frame in each video. These objects are then tracked throughout the video using a six dimensional $(x, y, z, r, g, b)$ particle filter, as shown in Fig. 8.

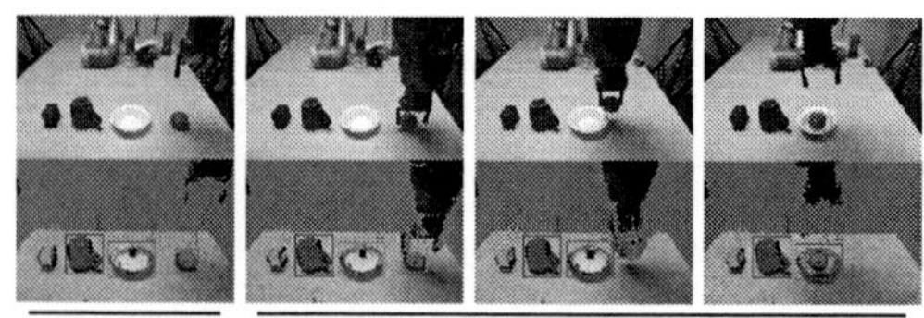


Figure 8: Example of a video sequence "place the orange in the bowl" when the objects are tracked using a particle filter. (Best viewed in colour.)

During the learning process, we use *atomic actions* (Fig. 2) to represent more **complex actions** in videos. For example a 'pick up' action is

---

[1] *Train Robots*: http://doi.org/10.5518/32

[2] Baxter dataset: http://doi.org/10.5518/110

represented with the sequence ($approach$, $grasp$, $lift$) as the robot approaches, grasps and lifts the object, while a 'drop' action is represented with just ($discard$) as the robot lets go of the object to fall down on the table.

We automatically detect **function words** by setting a threshold of $\sigma = .6$ on the Hungarian algorithm. Thus a word $w$ is considered a function word if it is not consistent with any cluster $f_j \in F$ by more than $60\%$ in the entire dataset. This threshold detects all function words.

In our **experiments**, we divide each dataset randomly into four equal parts, and perform four-fold cross validation, where we train on three folds and test on the fourth.

## 7.2 Evaluation

We evaluated the performance of our technique using two metrics: ($i$) the ability to correctly ground words to the learned visual concepts using $\Phi$, and ($ii$) the ability to correctly parse previously unseen natural language commands to produce correct RCL trees using the learned grammar $G$.

To better demonstrate our results in language grounding and grammar induction, we compare our technique with (1) a supervised system that learns from labelled data, and with (2) an unsupervised system that learns from unlabelled linguistic data. We consider our **baseline** as the performance of the unsupervised system, i.e. our joint language and vision technique should outperform the unsupervised system that learns from unlabelled linguistic inputs, otherwise there is no benefit of the additional vision component. On the other hand, an **upper bound** on performance is the results of the supervised system trained on human labelled (ground-truth) data.

## 7.3 Language Grounding Experiment

In this section, we evaluate the system's ability to acquire correct groundings for words from parallel pairs of short video clips and linguistic descriptions. The given task is to learn the partial function $\Phi : W \rightarrow C$ that maps words $w_i \in W$ to their corresponding clusters $c_j \in C$, e.g. the word 'red' should be mapped to the cluster colour-*red*.

The results for our language grounding experiment are shown in Fig. 9. Here, 'our-system' is compared against (1) the supervised semantic tagger (Fonseca and Rosa, 2013) that is trained on human labelled data, and (2) the unsupervised semantic tagger (Biemann, 2009) that is trained on unlabelled linguistic data. The results are

calculated based on the total number of correct tags/groundings assigned to each word in the test fold (four fold cross validation). Note that for the unsupervised system, the results are calculated based on its ability to cluster words that belong to the same category together, i.e. words that describe colours should be given a unique tag different to those that describe shapes, directions, etc. Also, we assign new words in the test fold (words that only exist in the test fold) with a *function word* tag.

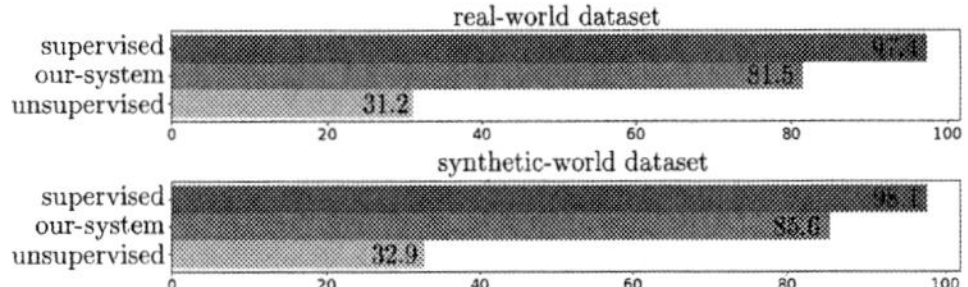

Figure 9: The grounding results of ($a$) supervised, ($b$) our system, and ($c$) unsupervised semantic taggers, on both datasets.

Our system is able to correctly ground ($85.6\%$) of the total words in the synthetic, and ($81.5\%$) in the real-world datasets, compared to only ($32.9\%$ and $31.2\%$ respectively) using the unsupervised system. This clearly shows that adding vision inputs produces more correct semantic representations for words, even though both systems use unlabelled data for learning. Detailed analysis of how the different techniques performed in each feature space is shown in Fig. 10. Note that $distance$ is not a feature in the synthetic dataset and therefore the corresponding row/column are left empty.

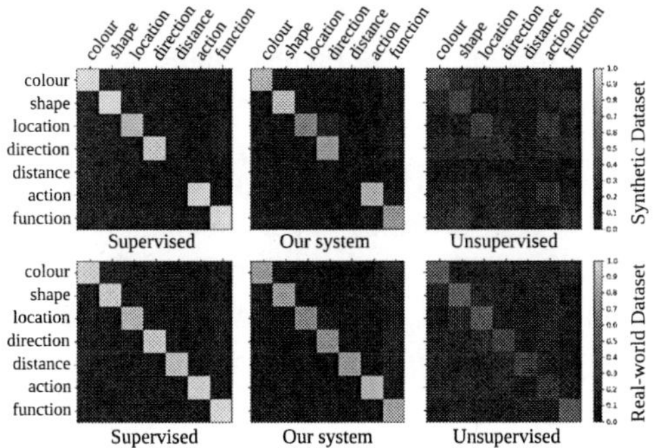

Figure 10: The grounding/tagging performance in each feature space for the three systems on both real-world and synthetic datasets.

## 7.4 Grammar Induction Experiment

In this experiment we test our system's ability to acquire correct grammar rules $G$ from pairs of video clips and unlabelled sentences (with no human-annotated RCL trees). The learned grammar $G$ is then used to parse new (previously unseen) natural language commands. We compare our technique with (1) a supervised parser (Abney, 1996) trained on labelled data, i.e. pairs of sentences and human-annotated RCL trees, and (2) an unsupervised parser (Ponvert *et al.* 2011) trained on unlabelled sentences, i.e. a corpus of sentences without RCL trees or semantic tags.

The results for ($a$) our approach, ($b$) the supervised parser, and ($c$) the unsupervised grammar induction systems on both datasets are shown in Fig. 11. The results were calculated based on the number of correctly parsed RCL trees from sentences in the test fold (in the four-fold cross validation). A score of 1 is given if the parsed sentence completely matches the human annotation, while a partial score in $(0, 1)$ is given if it partially matches the human annotation. The partial matching is computed by matching subtrees in the both trees divided by the total number of subtrees. For example, if a tree contains 10 subtrees and only 8 of which has a complete match in labels and links, then we give a score of 0.8 to this tree.

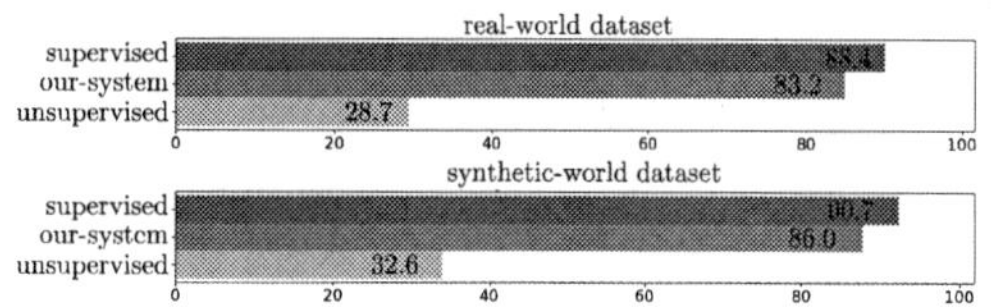


Figure 11: The grammar induction results for ($a$) supervised, ($b$) our system, and ($c$) unsupervised parsers on both real-world and synthetic datasets.

The results in Fig. 11 clearly show that our approach outperforms the unsupervised grammar induction system and achieves comparable results to the supervised system by learning from both language and vision as opposed to learning from language alone. The number of grammar rules generated differs between techniques: our approach generated (139 and 87) grammar rules from the synthetic and real-world datasets respectively, while the supervised system generated (182 and 114) and the unsupervised system generated (45 and 38) grammar rules, respectively.

## 8  Conclusion and Discussion

We present a novel technique to simultaneously learn the groundings of words and simple grammar rules of natural language. Our learning framework connects words from sentences to automatically-extracted visual clusters from videos to enable automatic generation of RCL trees, which is a key contribution of this paper. These trees act as an intermediary representation between the continuous perceptual space and the purely symbolic linguistic structures, thus provide robots with the ability to automatically learn about language, actions and perception. Our approach outperforms unsupervised techniques in both semantic tagging and grammar induction in learning from unlabelled data, and provides comparable performance to language-only supervised approaches.

Our approach suffers from two main limitations that hinder learning from longer videos (such as *YouTube* videos). First, it requires the videos and sentences to be temporally aligned beforehand, and second, it requires the feature spaces (e.g. *colours, shapes*, etc.) to be specified beforehand (though not their discretisation, which is learned). In order to allow learning from such data, our system should be able to learn from continuous, un-aligned videos and documents, and it should be able to generate new feature spaces to cope with new emerging concepts. We aim to address these limitations in future work.

## Acknowledgments

We acknowledge the financial support provided by EU FP7 project 600623 (STRANDS), and the the anonymous referees for their useful comments.

## References

Steven Abney. 1996. Partial parsing via finite-state cascades. *Natural Language Engineering* 2(04).

Jean-Baptiste Alayrac, Piotr Bojanowski, Nishant Agrawal, Josef Sivic, Ivan Laptev, and Simon Lacoste-Julien. 2016. Unsupervised learning from narrated instruction videos .

Chris Biemann. 2009. Unsupervised part-of-speech tagging in the large. *Research on Language and Computation* 7(2-4):101–135.

Joris Bleys, Martin Loetzsch, Michael Spranger, and Luc Steels. 2009. The grounded colour naming game. *Proceedings of Spoken Dialogue and Human-Robot Interaction Workshop at the RoMan 2009 Conference* .

Danushka Bollegala, Alsuhaibani Mohammed, Takanori Maehara, and Ken-ichi Kawarabayashi. 2015. Joint word representation learning using a corpus and a semantic lexicon. *arXiv preprint arXiv:1511.06438* .

Eugene Charniak. 1997. Statistical parsing with a context-free grammar and word statistics. *AAAI/IAAI* 2005(598-603):18.

David L Chen and Raymond J Mooney. 2011. Learning to interpret natural language navigation instructions from observations. In *AAAI*. volume 2.

Wanyun Cui, Xiyou Zhou, Hangyu Lin, Yanghua Xiao, Haixun Wang, Seung-won Hwang, and Wei Wang. 2016. Verb pattern: A probabilistic semantic representation on verbs .

Kais Dukes. 2013. Semantic annotation of robotic spatial commands. In *Language and Technology Conference (LTC)*.

Kais Dukes. 2014. Semeval-2014 task 6: Supervised semantic parsing of robotic spatial commands. *SemEval 2014* page 45.

Erick R Fonseca and Joao Luis G Rosa. 2013. A two-step convolutional neural network approach for semantic role labeling. In *Neural Networks (IJCNN), The 2013 International Joint Conference on*. IEEE.

Rein Houthooft, Cedric De Boom, Stijn Verstichel, Femke Ongenae, and Filip De Turck. 2016. Structured output prediction for semantic perception in autonomous vehicles. In *AAAI*. AAAI Press.

Eric H Huang, Richard Socher, Christopher D Manning, and Andrew Y Ng. 2012. Improving word representations via global context and multiple word prototypes. In *ACL*. ACL, pages 873–882.

Ulrich Klank, Dejan Pangercic, Radu Bogdan Rusu, and Michael Beetz. 2009. Real-time CAD Model Matching for Mobile Manipulation and Grasping. In *9th IEEE-RAS International Conference on Humanoid Robots*. Paris, France, pages 290–296.

Dan Klein and Christopher D Manning. 2002. A generative constituent-context model for improved grammar induction. In *ACL*. ACL, pages 128–135.

Harold W Kuhn. 1955. The Hungarian method for the assignment problem. *Naval research logistics quarterly* 2(1-2):83–97.

Jonathan Malmaud, Jonathan Huang, Vivek Rathod, Nick Johnston, Andrew Rabinovich, and Kevin Murphy. 2015. What's cookin'? interpreting cooking videos using text, speech and vision. *arXiv preprint arXiv:1503.01558* .

Cynthia Matuszek, Evan Herbst, Luke Zettlemoyer, and Dieter Fox. 2013. Learning to parse natural language commands to a robot control system. In *Experimental Robotics*. Springer, pages 403–415.

Dipendra K Misra, Jaeyong Sung, Kevin Lee, and Ashutosh Saxena. 2015. Tell me Dave: Context-sensitive grounding of natural language to manipulation instructions. *JAIR* page 0278364915602060.

Marius Muja and Matei Ciocarlie. 2013. tabletop object detector - ROS Wiki. `http://www.ros.org/wiki/tabletopobjectdetector`.

Elias Ponvert, Jason Baldridge, and Katrin Erk. 2011. Simple unsupervised grammar induction from raw text with cascaded finite state models. In *Proceedings of the 49th Annual Meeting of the ACL: Human Language Technologies*. ACL, Oregon, USA.

Deb Roy, Bernt Schiele, and Alex Pentland. 1999. Learning Audio-Visual Associations using Mutual Information. In *Integration of Speech and Image Understanding, 1999. Proceedings*. IEEE.

Hinrich Schütze. 1998. Automatic word sense discrimination. *Computational linguistics* 24(1):97–123.

Lanbo She, Shaohua Yang, Yu Cheng, Yunyi Jia, Joyce Y Chai, and Ning Xi. 2014. Back to the blocks world: Learning new actions through situated human-robot dialogue. In *15th Annual Meeting of the Special Interest Group on Discourse and Dialogue*. volume 89.

Jeffrey Mark Siskind. 1996. A Computational Study of Cross-Situational Techniques for Learning Word-to-Meaning Mappings. *Cognition* 61(1):39–91.

Noah A Smith and Jason Eisner. 2005. Contrastive estimation: Training log-linear models on unlabeled data. In *Proceedings of the 43rd Annual Meeting on Association for Computational Linguistics*. ACL.

Benjamin Snyder, Tahira Naseem, and Regina Barzilay. 2009. Unsupervised multilingual grammar induction. In *Proceedings of the Joint Conference of the 47th Annual Meeting of ACL*. ACL, pages 73–81.

Anders Søgaard. 2012. Unsupervised dependency parsing without training. *Natural Language Engineering* 18(02):187–203.

Karen Sparck Jones. 1972. A statistical interpretation of term specificity and its application in retrieval. *Journal of documentation* 28(1):11–21.

Michael Spranger and Luc Steels. 2015. Co-acquisition of syntax and semantics - an investigation in spatial language. In Qiang Yang and Michael Wooldridge, editors, *IJCAI'15*, AAAI Press, Palo Alto, US, pages 1909–1905.

Luc Steels. 2001. Language Games for Autonomous Robots. *Intelligent Systems, IEEE* 16(5):16–22.

Luc Steels and Frederic Kaplan. 2002. Aibo's First Words: The Social Learning of Language and Meaning. *Evolution of Communication* 4(1):3–32.

Stefanie A Tellex, Thomas Fleming Kollar, Steven R Dickerson, Matthew R Walter, Ashis Banerjee, Seth Teller, and Nicholas Roy. 2011. Understanding natural language commands for robotic navigation and mobile manipulation .

# Communication with Robots using Multilayer Recurrent Networks

**Bedřich Pišl** and **David Mareček**
Charles University,
Faculty of Mathematics and Physics,
Institute of Formal and Applied Linguistics
`bedapisl@gmail.com, marecek@ufal.mff.cuni.cz`

## Abstract

In this paper, we describe an improvement on the task of giving instructions to robots in a simulated block world using unrestricted natural language commands.

## 1 Introduction

Many of the recent methods for interpreting natural language commands are based mainly on semantic parsers and hand designed rules. This is often due to small datasets, such as Robot Commands Treebank (Dukes, 2013) or datasets by MacMahon et al. (2006) or Han and Schlangen (2017).

Tellex et al. (2011) and Walter et al. (2015) present usage of such systems in real world. They developed a robotic forklift which is able to understand simple natural language commands. For training, they created small dataset by manually annotating the data from Amazon Mechanical Turk. Their model is based on probabilistic graphical models invented specifically for this task.

The first approach using neural networks is proposed by Bisk et al. (2016b), who describe and compare several neural models for understanding natural language commands. Their dataset (Bisk et al., 2016a) contains simulated world with square blocks and actions descriptions in English (see Figure 1). Since the actions are always shifts of single block to some location, they divide the task into two: predicting which block should be moved and where. They call these tasks source and target predictions. With their best model, they reach 98% accuracy for source prediction and 0.98 average distance between correct and predicted location for target.

The world is represented by $x$ and $y$ coordinates of 20 blocks. Each block has a digit or logo of a company for easy identification. There are 16,767

Figure 1: Visualisation of command "Move Nvidia block to the left of HP block" in our world.

commands in the dataset, divided into train, development, and test set. The commands were written by people using Amazon Mechanical Turk and therefore contains many typos and other errors.

In this paper, we propose several models solving this task and report improvement compared to the previous work by Bisk et al. (2016b).

## 2 Models

### 2.1 Data preprocessing

For tokenization of commands we use simple rule based system. Because of the typos we use Hunspell[1], which is a widely used spell checker. Finally to prevent overfitting of neural models we replace all tokens with less than 4 occurrences in training data with special token representing unknown word.

### 2.2 Benchmark model

To be able to measure the impact of the RNN based models, we first introduce a simple rule-based benchmark. The benchmark searches for block numbers (*1, 2, ..., 20, one, two, ... twenty*), logo names (*adidas, bmw, burger, king, ...*),[2] and directions (*west, north, east, south, left, above, right, below*) in the commands.

---

[1] http://hunspell.github.io/

[2] If the logo name contains more words (e.g. *burger king*), this model search for each part of the word and for concatenation of both words (e.g. *burger, king*, and *burgerking*).



*Proceedings of the First Workshop on Language Grounding for Robotics*, pages 44–48,
Vancouver, Canada, July 30 - August 4, 2017. ©2017 Association for Computational Linguistics

For predicting the source (which block should be moved), the model predicts the block corresponding to the first word in the sentence denoting a block. For predicting the target location (where the source block should be moved), the model predicts position of the last word describing block If there exist words describing directions, the last one is chosen and the position is changed by one in the direction corresponding to the word.

For example, in the command

> *Put the **UPS** block in the same column as the **Texaco** block, and one row **below** the **Twitter** block.*

the benchmark model finds three words describing blocks (*UPS, Texaco,* and *Twitter*) and the word *below* describing direction. The block word (*UPS*) is predicted as source. As the target location, the banchmark model chooses the current location of *Twitter* block (the last block word) moved one tile down, because of the *below* word.

### 2.3 Neural model with world on the input

Our first neural model is relatively straightforward. Word embedding vectors representing the tokenized command are given to a bidirectional LSTM recurrent layer (Hochreiter and Schmidhuber, 1997). The last two states of both directions are concatenated together with the world representation (2 coordinates for each of the 20 blocks). and fed into single feed forward layer with linear activation function. For predicting source, this layer has dimension 20 and its outputs are then used as logits to determine the source block. For predicting location the last feed forward layer has dimension two and its outputs are directly interpreted as predicted target location.

### 2.4 Predicting reference and relative position

Our second model is similar to the one proposed by Bisk et al. (2016b).

It does not predict directly the target location, but a meaning representation of the command, which is then interpreted based on the world state to get the final predicted target location. Our representation is composed of 20 weights representing how much each block is used as a reference, and 2-dimensional vector representing the relative position from the reference block. Let $w = (w_1, w_2, ...w_{20})^T$ represent the weights of individual reference blocks, $d = (d_1, d_2)^T$ represent

the relative position and

$$S = \begin{pmatrix} s_{1,1} & s_{1,2} & \cdots & s_{1,20} \\ s_{2,1} & s_{2,2} & \cdots & s_{2,20} \end{pmatrix}$$

be the state of world, where $s_{1,i}$ and $s_{2,i}$ are $x$ and $y$ coordinates of the $i$-th block The final target location $l \in R^2$ is then computed as $l = Sw + d$.

In most commands, the target is described in one of the following ways:

1. By reference and direction: *Move BMW above Adidas*
2. By reference, distance and direction: *Move BMW 3 spaces above Adidas*
3. By absolute target: *Move BMW to the middle of bottom edge of the table*
4. By direction relative to source: *Move BMW 3 spaces down*
5. By two references: *Move BMW between Adidas and UPS*

This representation is able to capture the meaning of all of these. For example, the command 1 can be represented as $w = (1, 0, 0, ..., 0)^T, d = (0, 1)^T$, the command 5 as $w = (0.5, 0, 0, ..., 0, 0, 0.5)^T, d = (0, 0)^{T}.$[3]

The tokenized one-hot encoded command is given to a bidirectional LSTM recurrent layer, the two last states are concatenated and fed into two parallel feed-forward layers. The first one has 20 dimensions and outputs the weights $w$ of references, the second one is 2-dimensional and outputs the relative position $d$. The target location is then computed from these.

### 2.5 Using recurrent output layers

We also tested a variant of the previous architecture in which the feed-forward output layers are substituted by recurrent 128-dimensional LSTM layers. The new architecture is shown in Figure 2.

We also tried similar models for predicting the source blocks. They have bidirectional recurrent layer, followed by single output layer, which is feed-forward for one model and recurrent 64-dimensional LSTM for the other one.

## 3 Results

The experiment results are compared in Table 3. We report improvement over the previous results for both source and target location predictions. For source prediction the network without world

---

[3]The Adidas block has weight $w_1$ and the UPS block is the last with weight $w_{20}$.

Figure 2: Final target prediction network, which uses recurrent layer instead of feed-forward one.

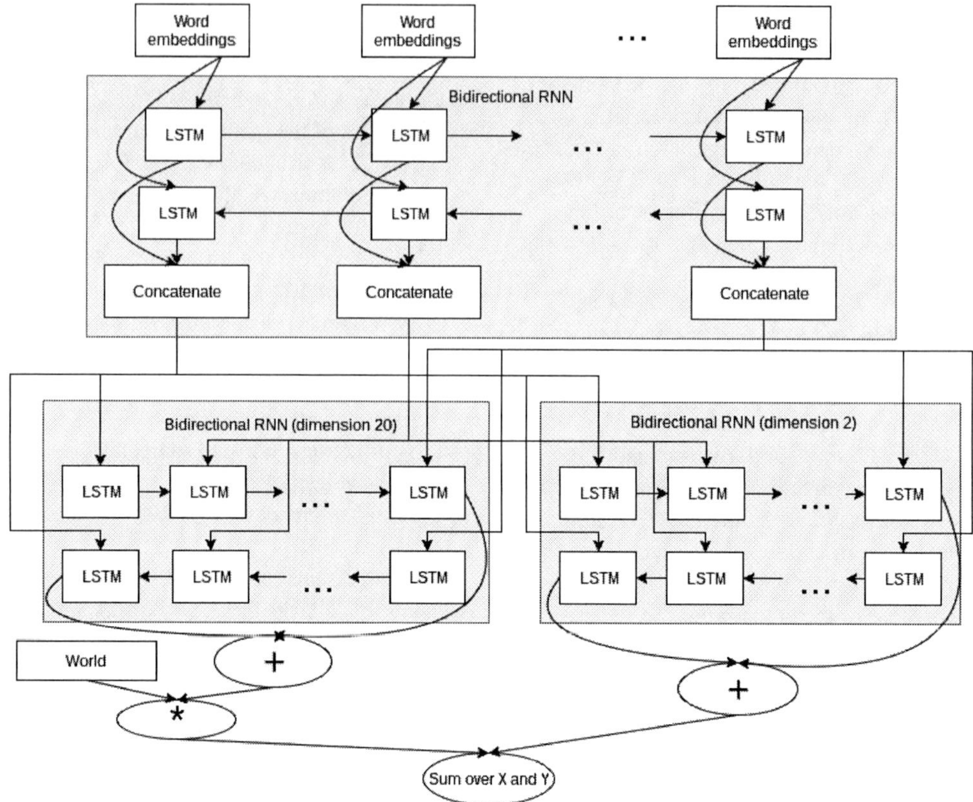

on the input and with feed-forward output layer achieves accuracy 98.8%.This is better than the best model of Bisk et al. (2016b), who reported 98% accuracy. The improvement is mainly caused by preprocessing data with spell checker and better hyperparameter selection. Without using spell checker our model has accuracy 98.3%.

As for the target location prediction, our best model has average distance of 0.72 between predicted and correct target location. This is an improvement over both rule based benchmark with 1.54 and the best model reported by Bisk et al. (2016b), who had 0.98. The median distance is 0.04 which is much better than their comparable End-To-End model with median distance 0.53. In 65.8% of test instances the distance of our model is less than 0.5, which might be considered a distinctive line between good and bad prediction.

## 4   Error analysis and discussion

We manually analyzed bad predictions of our best model. As for the source block prediction, there were only 18 mistakes made on the devset:

1. The two-sentence command (7 mistakes). In the first sentence, it looks like the first mentioned block is the source, but the second sentence states otherwise.[4] *"The McDonald's tile should be to the right of the BMW tile. Move BMW."*

2. Block switching (3 mistakes): *"The 16 and 17 block moved down a little but switched places."*

3. Commands with typos (3 mistakes): *"Slide block the to the space above block 4"* (Note that the third word here should be *three*.)

4. Commands including a sequence (2 mistakes): *"Continue 13, 14, 15…"*

5. Grounding error (2 mistakes), see Table 1.

6. Annotation error (once, not a mistake).

Major improvement of source accuracy may be achieved by solving the problem where second sentence changes the meaning of the first one. However, there are no similar commands in the training data, so it is hard to come with solution.

---

| Mistake type | # | Description & Example |
|---|---|---|
| More reference blocks | 31 | Target location is described using two or more reference blocks. *Place block 12 on the same horizontal plane as block 9, and one column left of block 14.* |
| Source same as reference block | 11 | Model mistakes source for reference. Typically, the last block mentioned in the sentence is source. *Move block 10 above block 11, evenly aligned with 10 and slightly separated from the top edge of 10.* |
| Annotation error | 11 | Command does not make sense or does not describe the correct action. *Move Pepsi to the slight southwest until it's north of Pepsi.* |
| Missing reference block | 9 | No reference block in the command. *Move the Stella block down to the very bottom of the square.* |
| Large direction | 9 | Distance between reference and correct location is more than 1 block. *move the texaco block 5 block lengths above the BMW block* |
| Grounding error | 8 | Unusual description of blocks and typos in block names. *Put the block that looks like a taurus symbol just above the bird.* |
| Learning mistake | 8 | Relatively simple example, yet still bad prediction. *Block 4 should be moved almost straight down until it is resting on block 5.* |
| Others | 13 | |

Table 1: The worst predictions analysis: Probable reasons behind bad predictions in 100 worst instances of the development set. *References* are the blocks which are used in the command for describing the target location.

| | Source | Target |
|---|---|---|
| Random baseline | 5.4% | 6.12 |
| Middle baseline | 5.4% | 3.46 |
| Rule-based benchmark | 96.3% | 1.54 |
| Bisk et al. (2016b) | 98% | 0.98 |
| World as input | 98.5% | 3.05 |
| Feed-forward output layer | 98.8% | 1.07 |
| Recurrent output layer | 98.5% | 0.72 |

Table 2: Results comparison. In *random* and *middle* baselines, the randomly chosen block is placed on random position or in the middle of the board.

Similarly, the word *switch* appears only once in the training set.

Overall we think that for source prediction we reached the limitations given by the dataset we are using and without usage of another data it is very hard to get significant improvements.

For target prediction we divide 100 worst predictions into categories, which can be seen in Table 1.

11 out of the 100 worst predictions are bad because the commands does not make sense. But also in many other commands the target location is not described precisely, so the overall impact of inaccurate commands is in our opinion bigger and it also influences the training of models.

The other problem categories except of Learning mistake have similar underlying cause. The sentence structure is unusual and does not appear in the training data very often. Also in some cases such as the More references category the sentences are more complicated.

But even though these sentences are challenging and the model makes mistakes in them relatively often, it works well for majority of these sentences. Thus we find out that our proposed sentence representation is in practice capable of representing almost all sentences in the dataset.

## 5   Conclusion

We presented four different architectures of neural networks for solving the task of robot communication on dataset by Bisk et al. (2016a). Our last model surpassed the previous reported results and reached accuracy of 98.8% for source prediction and 0.72 average distance between predicted and correct target location. We find out that our model is capable of understanding wide variety of commands in natural language and make mistakes mostly in sentences with features, which are badly or not at all represented in the training data.

# References

Yonatan Bisk, Daniel Marcu, and William Wong. 2016a. Towards a dataset for human computer communication via grounded language acquisition. In *Workshops at the Thirtieth AAAI Conference on Artificial Intelligence*.

Yonatan Bisk, Deniz Yuret, and Daniel Marcu. 2016b. Natural language communication with robots. In *Proceedings of NAACL-HLT*. pages 751–761.

Kais Dukes. 2013. Semantic annotation of robotic spatial commands. In *Language and Technology Conference (LTC)*.

Ting Han and David Schlangen. 2017. Grounding language by continuous observation of instruction following. *EACL 2017* page 491.

Sepp Hochreiter and Jürgen Schmidhuber. 1997. Long short-term memory. *Neural Comput.* 9(8):1735–1780.

Matt MacMahon, Brian Stankiewicz, and Benjamin Kuipers. 2006. Walk the talk: connecting language, knowledge, and action in route instructions, proceedings of the 21st national conference on artificial intelligence.

Stefanie A Tellex, Thomas Fleming Kollar, Steven R Dickerson, Matthew R Walter, Ashis Banerjee, Seth Teller, and Nicholas Roy. 2011. Understanding natural language commands for robotic navigation and mobile manipulation .

Matthew R Walter, Matthew Antone, Ekapol Chuangsuwanich, Andrew Correa, Randall Davis, Luke Fletcher, Emilio Frazzoli, Yuli Friedman, James Glass, Jonathan P How, et al. 2015. A situationally aware voice-commandable robotic forklift working alongside people in unstructured outdoor environments. *Journal of Field Robotics* 32(4):590–628.

# Grounding Symbols in Multi-Modal Instructions

**Yordan Hristov, Svetlin Penkov, Alex Lascarides and Subramanian Ramamoorthy**
School of Informatics
The University of Edinburgh
{yordan.hristov@, sv.penkov@, alex@inf., s.ramamoorthy@}ed.ac.uk

## Abstract

As robots begin to cohabit with humans
in semi-structured environments, the need
arises to understand instructions involving rich variability—for instance, learning
to ground symbols in the physical world.
Realistically, this task must cope with
small datasets consisting of a particular
users' contextual assignment of meaning
to terms. We present a method for processing a raw stream of cross-modal input—
i.e., linguistic instructions, visual perception of a scene and a concurrent trace
of 3D eye tracking fixations—to produce
the segmentation of objects with a cor-
respondent association to high-level concepts. To test our framework we present
experiments in a table-top object manipulation scenario. Our results show our
model learns the user's notion of colour
and shape from a small number of physical demonstrations, generalising to identifying physical referents for novel combinations of the words.

## 1 Introduction

Effective and efficient human-robot collaboration
requires robots to interpret ambiguous instructions
and concepts within a particular context, communicated to them in a manner that feels natural and
unobtrusive to the human participant in the interaction. Specifically, the robot must be able to:

- Understand natural language instructions,
  which might be ambiguous in form and
  meaning.

- Ground symbols occurring in these instructions within the surrounding physical world.

Figure 1: Combining natural language input with
eye tracking data allows for the dynamic labelling
of images from the environment with symbols.
The images are used in order to learn the meaning
of language constituents and then ground them in
the physical world.

- Conceptually differentiate between instances
  of those symbolic terms, based on features
  pertaining to their grounded instantiation,
  e.g. shapes and colours of the objects.

Being able to relate abstract symbols to observations with physical properties in the real world is
known as the physical symbol grounding problem
(Vogt, 2002); which is recognised as being one of
the main challenges for human-robot interaction
and constitutes the focus of this paper.

There is increasing recognition that the meaning of natural language words derives from how
they manifest themselves across multiple modalities. Researchers have actively studied this problem from a multitude of perspectives. This includes works that explore the ability of agents to
interpret natural language instructions with respect



*Proceedings of the First Workshop on Language Grounding for Robotics*, pages 49–57,
Vancouver, Canada, July 30 - August 4, 2017. ©2017 Association for Computational Linguistics

to a previously annotated semantic map (Matuszek et al., 2013) or fuse high-level natural language inputs with low-level sensory observations in order to produce a semantic map (Walter et al., 2014). Matuszek et al. (2014); Eldon et al. (2016) and Kollar et al. (2013) tackle learning symbol grounding in language commands combined with gesture input in a table-top scenario. However, all these approaches depend on having *predefined* specifications of different concepts in the environment: they either assume a pre-annotated semantic map with respect to which they ground the linguistic input or have an offline trained symbol classifier that decides whether a detected object can be labelled with a specific symbol; e.g. colour and shape in (Matuszek et al., 2014). Thus in order to deploy such a system, one should have access to an already trained classifier for every anticipated symbol, prior to any user interaction.

Multi-modal learning algorithms based on deep neural networks are also popular for grounding natural language instructions to the shared physical environment (Srivastava and Salakhutdinov, 2012; Ngiam et al., 2011). But the majority of these algorithms depend crucially on large and pre-labelled datasets, and the challenge is in collecting these large-scale labelled datasets so that they not only capture the variability in language but also manage to represent the nuances (especially across multiple high-bandwidth modalities, such as vision and eye-tracking) of inter-personal variability in assignment of meaning (e.g., what one person calls *mustard* another might call *yellow*), which we claim is a key attribute of free-form linguistic instructions in human-robot interaction applications. If a previously unseen instruction/visual observation is presented to these systems, they might fail to ground or recognize them in the way that the user might have intended in that specific setting. Tobin et al. (2017) potentially bypasses the need to collect a big dataset by demonstrating that a model trained in simulation can be successfully deployed on a robot in the real world. However, the problem is then shifted to generating task-specific training data in a simulator which approximates the real world well enough.

A proposed alternative to this off-line learning approach is to interactively teach an embodied agent about its surrounding world, assuming limited prior knowledge. Al-Omari et al. (2016) demonstrates a model for incrementally learning the visual representation of words, but relies on temporally aligned videos with corresponding annotated natural language inputs. Parde et al. (2015) and Thomason et al. (2016) represent the online concept learning problem as a variation of the interactive "I Spy" game. However, these approaches assume an initial learning/exploratory phase in the world and extracted features are used as training data for all concept models associated with an object.

Penkov et al. (2017) introduce a method called GLIDE (see §2.2 for details), which successfully teaches agents how to map abstract instructions, represented as a LISP-like program, into their physical equivalents in the world. Our work builds on this method: it uses it to achieve *natural language* symbol grounding, as a by-product of user interaction in a task-oriented scenario. Our approach achieves the following:

- It maps natural language instructions to a planned behaviour, such as in a robotic manipulation domain; in so doing it supports a communication medium that human users find natural.

- It learns symbol grounding by exploiting the concept of *intersective modification* (Morzycki, 2013) —i.e., an object can be labelled with more than one symbol. The meaning of the symbols is learned with respect to the observed features of the instances of the object.

In our work the agent assumes some prior knowledge about the world in the form of low-level features that it can extract from objects in the visual input—e.g. intensities in the primary colour channels and areas of pixel patches of any specific colour. On top of this, we learn classifiers for performing symbol grounding. Each symbol has a probabilistic model which is fit to a subset of the extracted (visual) features. When a new instruction is received, the classifier for each symbol makes a decision regarding the object in the world (and their respective features) to which the symbol may be grounded. Crucially, the data from which these classifiers are learned is collected from demonstrations at 'run time' and not prior to the specific human-robot interaction. Images of objects are extracted from the high-frequency eye tracking and video streams, while symbols that refer to these objects in the images are extracted from the parsed natural language instructions—see Figure 1. Through cross-modal instructions,

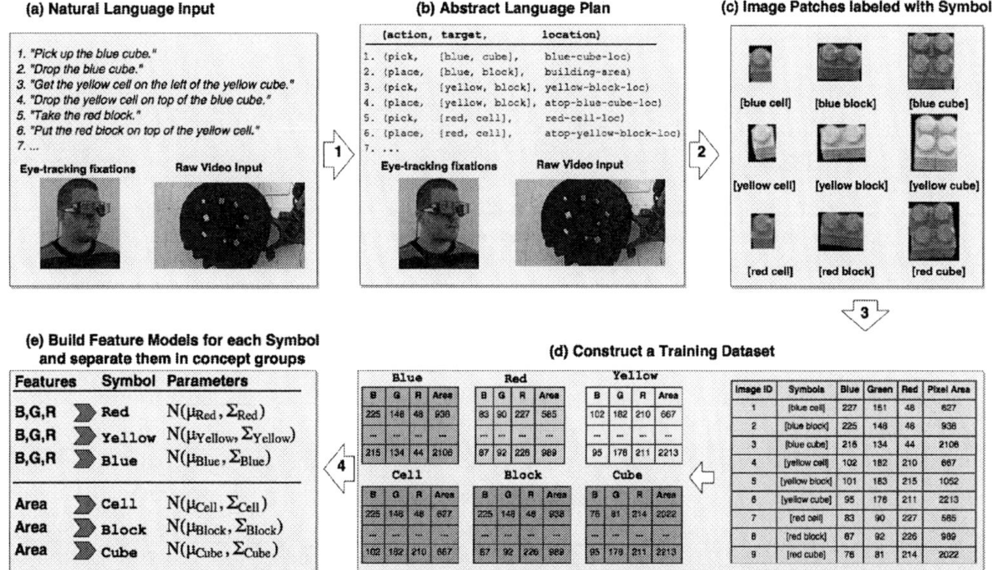


Figure 2: Overview of the full system pipeline. Input to the system are natural language instructions, together with eye-tracking fixations and a camera view of the world from above **(a)** Natural language instructions are deterministically parsed to an abstract plan language **(b)** Using the abstract plan, a set of labelled image patches is produced from the eye-tracking and video data **(c)** Observable predefined features are extracted from the image patches **(d)** Each symbol is grounded to a subset of observable features **(e)**

the human participant is simultaneously teaching the robot how to execute a task and what properties the surrounding objects must have for that execution to be successful. For instance, while observing how to make a fruit salad in a kitchen, apart from learning the sequence of steps, the system would also gain an initial approximation of the visual appearance of different pieces of fruit and their associated natural language symbols.

## 2 Methods

Figure 2 depicts the architecture of the overall system. It consists of an end-to-end process, from raw linguistic and video inputs on the one hand to learned meanings of symbols that in turn are conceptually grouped: i.e., a symbol can correspond either to an object in the real world, or to a property of an object. The rest of the section is organized in the following fashion - each subsection corresponds to a numbered transition (1 to 4) indicated in Figure 2.

### 2.1 Natural Language Semantic Parsing

The task of the semantic parser is to map natural language requests into instructions represented in an abstract form. The abstract form we use is a list of tuples with the format `(action target location)` (Figure 2b), where `action` corresponds to an element from a predefined set $\mathcal{A}$, `target` corresponds to a list of terms that describe an object in the world and `location` corresponds to a single symbol denoting a physical location in the environment.

The narration of the plan execution by the human comprises one sentence per abstract instruction. Therefore, given a plan description, our semantic parser finds a mapping from each sentence to a corresponding instruction as defined by our abstract plan language.

Elementary Dependency Structures (EDS) (Oepen et al., 2004), which are output by parsing the sentence with the wide-coverage English Resource Grammar (Flickinger et al., 2014), are used as an intermediate step in this mapping procedure. EDS are given as dependency graphs (Figure 3) and are a variable-free reduced form of the full Minimal Recursion Semantics (MRS) (Copestake et al., 2005) representation of the natural language input. Given EDS for a particular sentence, parsing proceeds in two steps:

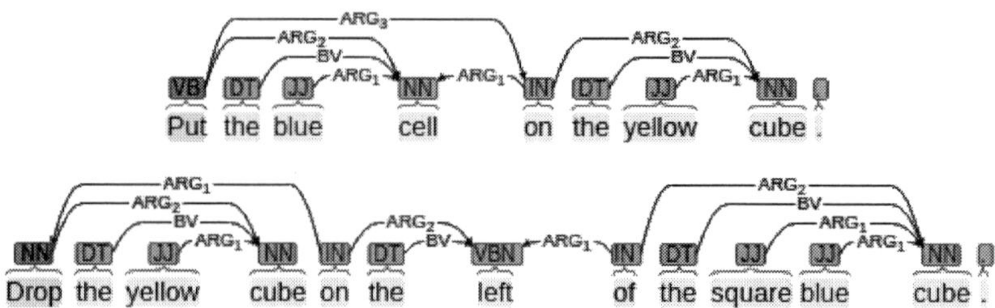

Figure 3: Example of dependency graphs for input sentences. Red labels denote the top graph nodes.

- The graph is extracted from nodes and their respective edges, cleaning up nodes and edges that do not contribute directly to the meaning of the instruction—i.e., anything that is not a verb, adjective, preposition or a noun;

- The processed graph is recursively traversed until all nodes have been visited at least once.

Knowing the format of our abstract plan language, we know that `action` would always correspond to the verb in the input sentence, `target` would correspond to a noun phrase and `location` would correspond to a prepositional phrase (i.e., a combination of a preposition and noun phrase). For us, the noun phrases all consist of a noun, complemented by a possibly empty list of adjectives. Extracting the action is straightforward since the top node in the EDS always corresponds to the verb in the sentence; see Figure 3. The extracted action is then passed through a predefined rule-based filter which assigns it one of the values from $\mathcal{A}$: e.g. *pick, grab, take, get* would all be interpreted as `pick`.

The `target` entry can be extracted by identifying noun node in the EDS that's connected to the verb node. Once such a noun is found, one can identify its connections to any adjective nodes—this gives a full list of symbols that define the object referenced by the `target`.

The `location` entry can be extracted by searching for preposition nodes in the EDS that are connected to the verb node. If there is no such node, then the `location` is constructed directly from the target by concatenating its labels - e.g. for a blue cube the `location` would be the symbol `blue-cube-location`. Extracting the `location` from a prepositional phrase is less constrained since different verbs can be related to spatial prepositions in varied ways— either the preposition node has an edge connecting it to the verb node or vice versa. Once a prepositional node is visited, we proceed by recursively exploring any chains of interconnected nouns, prepositions, and adjectives. The recursion calls for backtracking whenever a node is reached with no unvisited incoming or outgoing edges: e.g., node `cube` on Figure 3 (bottom). For example, the symbol `on-left-of-cube` is produced for `location` for the bottom sentence in Figure 3.

In this way, the result of parsing is a sequence of abstract instructions—i.e., an abstract plan— together with a *symbol set S*, containing all symbols which are part of any `target` entry. At this point, the symbols are still not actually grounded in the real world. Together with the raw video feed and the eye-tracking fixations, the abstract plan becomes an input to GLIDE (Penkov et al., 2017).

## 2.2 Grounding and Learning Instances through Demonstration and Eye tracking (GLIDE)

Penkov et al. (2017) introduce a framework for Grounding and Learning Instances through Demonstration and Eye tracking (GLIDE). In this framework, *fixation programs* are represented in terms of fixation traces obtained during task demonstrations combined with a high-level plan. Through probabilistic inference over fixation programs, it becomes possible to infer latent properties of the environment and determine locations in the environment which correspond to each instruction in an input abstract plan that conforms to the format discussed above - (`action target location`).

### 2.2.1  3D Eye-Tracking

Mobile eye trackers provide fixation information in pixel coordinates corresponding to locations in the image of a first person view camera. In order to utilise information from multiple input sensors,

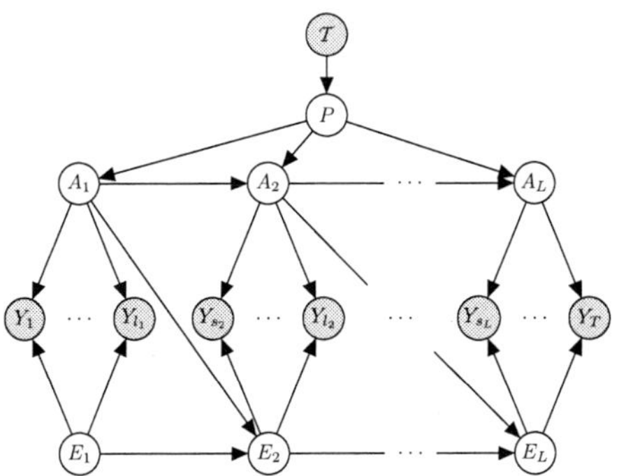

Figure 4: The proposed probabilistic model for physical symbol grounding is based on the idea that "task and context determine where you look" (Rothkopf et al., 2007).

an additional camera may be attached to augment an eye-tracker, by running mono camera SLAM algorithm in the background—ORBSLAM (Mur-Artal et al., 2015). The SLAM algorithm provides 6D pose of the eye tracking glasses within the world frame; this allows for the fixation locations to be projected into the 3D world by ray casting and finding intersections with a 3D model of the environment. As a result, fixations can be represented as 3D locations, enabling the projection of fixations in the frame of any sensor in the environment.

### 2.2.2 Model and Location Inference

In order to solve the problem of symbol grounding, inference is performed using a generative probabilistic model, which is shown in Figure 4. The sequence of fixations $Y_1 : Y_T$ depend both on the current environment state $E$ and the action being executed $A$. Each action is part of the plan $P$ which is determined by the task being demonstrated $T$. The sequence of fixations is observed and interpreted with respect to a task that is by this stage already known, while the state of the environment and the current action are unknown. The main inference task is to determine the structure of the model and assign each fixation to the action that is its cause. A crucial feature of this procedure is the fact, deriving from human sensorimotor behaviour, that the distribution of fixations is different when a person is attending to the execution of an action compared to periods of transitions between actions in the plan. By utilising this property and using samples from a Dirichlet distribution to describe these transition points, GLIDE is able to infer the correct partitioning of

the fixation sequence. This information allows us to localise each item of interest in the environment and extract labelled sensory signals from the environment. A complete description of this model and inference procedure can be found in (Penkov et al., 2017).

### 2.3 Feature Extraction

The parser produces a set of symbols $S$ and GLIDE produces a set of image patches $I$, each of which is labelled with a subset of symbols from $S$. We proceed by extracting a number of features, drawn from a pre-existing set $F$. The features are derived from the objects in the image patches, after removing the background. This is achieved through a standard background subtraction method—we know that the majority of the image will be occupied by a solid object with a uniform colour, anything else is a background. For instance, in the image patches in Figure 2 (c), the objects are the colourful blocks and the background is the black strips around them. Images containing only or predominantly background are considered noise in the dataset and are discarded. For each symbol $s$ we group the extracted features from each image labelled with $s$ resulting in $S$ lists of $M_s$ tuples with $F$ entries in each tuple, where $M_s$ is the number of images being labelled with $s$; see Figure 2 (d, left). The data for each feature is normalized to fall between 0 and 1.

---

**Algorithm 1:** Symbol Meaning Learning

**Input:** $\sigma_{thresh}$
**Data:** $I, S, F$
**Output:** $K = \{(\mu_1, \Sigma_1), \ldots, (\mu_S, \Sigma_S)\}$,
$\quad\quad\quad C = \{(F^s_{invar} : s), \ldots (F^S_{invar} : S)\}$

1   $Data \leftarrow [s_1 : \{\}, \ldots, s_S : \{\}]$;
2   **for** *image i in I* **do**
3     $symbols_i \leftarrow GetSymbols(i)$;
4     $features_i \leftarrow ExtractFeatures(i)$;
5     **for** *symbol in $symbols_i$* **do**
6       Append $features_i$ to $Data[symbol]$;
7   **for** *s in S* **do**
8     $K_s \leftarrow FitNormal(Data[s])$;
9     $K_s \leftarrow CleanNoise(Data[s])$;
10    $F^s_{invar} \leftarrow FindInvFeat(K_s, \sigma_{thresh})$;
11    $K_s \leftarrow RefitNormal(K_s, F^s_{invar})$;
12    Append $(F^s_{invar} : K_s)$ to $C$;

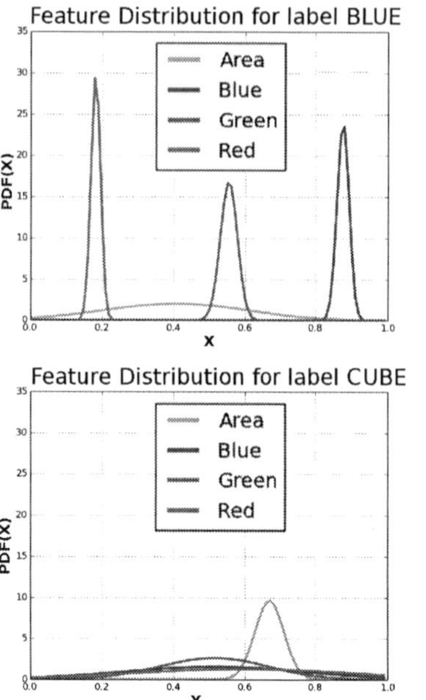


Figure 5: Example of feature distributions for blue (top) and cube (bottom) symbols.

### 2.4 Symbol Meaning Learning

For each symbol $s \in S$ and each feature $f \in F$ we fit a 1-D Normal distribution resulting in a new list of tuples with size $F$ - $s_j$ : $[(\mu_{f_1}^{s_j}, \sigma_{f_1}^{s_j}), \ldots, (\mu_{f_F}^{s_j}, \sigma_{f_F}^{s_j})]$ for the $j^{th}$ symbol. Taking into account that the object location process in GLIDE could still produce noisy results—i.e., the label of an image can be associated with the wrong symbol—we process our distributions to refit them to data that falls within two standard deviations from the means of the original distributions. We are then seeking observed features $f$ that are invariant with respect to each token use of a specific symbol $s$ within the user instructions so far—i.e. their distributions are 'narrow' and with variance below a predefined threshold $\sigma_{thresh}$ (see Figure 5). If we have a set of images that are of blue objects with different shapes and we extract a set of features from them, we would expect that features with lower variation (e.g. RGB channels as opposed to area) would explain the colour blue better than features with more variation (i.e. pixel area).

In the last step, we construct a set of the invariant features from the discovered narrow distributions for a given symbol $l$ - $(F_{invar}^s)$ - and say that this set characterizes the symbol. The parameters for the symbol are the concatenation of the means

of the features from $(F_{invar}^l)$ into a mean vector and the concatenation of the variances into a diagonal covariance matrix. The resultant mean vector and covariance matrix are later used for inference when shown a new set of images.

## 3 Experiments

We now present results from initial experiments based on the framework in Figure 2. We focus our explanation on steps 3 and 4 in that figure, as these are the pertinent and novel elements introduced here. The input data for Figure 2 (c) is derived from the process already well described in (Penkov et al., 2017).

### 3.1 Dataset

For our experiments we used a total of six symbols defining $S$: 3 for colour (red, blue and yellow); and 3 for shape (cell, block, cube). We used four extracted features for $F$: R, G, B values and pixel area. The objects used were construction blocks that can be stacked together and images of them were gathered in a tabletop robotic manipulation setup (see Figure 2 (a)). Based on the empirical statistics of the recognition process in (Penkov et al., 2017), our input dataset to the Symbol Meaning Learning algorithm consists of 75% correctly annotated and 25% mislabelled images. The total training dataset comprised of approximately 2000 labelled image patches, each of which is labelled with two symbols—e.g. blue cell, red block, yellow cube, etc.

The additional test set was designed in two parts: one that would test colour recognition and one that would test shape recognition. Overall, 48 objects were presented to the algorithm where the features for each object would fall into one of the following categories:

- Previously seen features (Figure 6 (left))

- Previously unseen features, close to the features of the training data (Figure 6 (middle))

- Previously unseen features, not close to the features of the training data (Figure 6 (right))

### 3.2 Experimental Set up

Inference over new images is performed by thresholding the probability density function (PDF) values from the model parameters for each symbol. The idea is to test how well the algorithm can differentiate the learned concepts with slight variations from concepts it has not seen before: e.g.

Figure 6: Variations in the objects from the test set for colour (top half) and shape (bottom half)

Table 1: Confusion matrix for colour symbols

|  | Red | Yellow | Blue | Unknown |
|---|---|---|---|---|
| **Red** | **12** | 1 | 0 | 0 |
| **Yellow** | 0 | **12** | 0 | 0 |
| **Blue** | 0 | 0 | **13** | 1 |
| **Unknown** | 0 | 1 | 0 | **13** |

Table 2: Confusion matrix for shape symbols

|  | Cell | Block | Cube | Unknown |
|---|---|---|---|---|
| **Cell** | **8** | 1 | 0 | 0 |
| **Block** | 1 | **7** | 0 | 0 |
| **Cube** | 0 | 3 | **8** | 0 |
| **Unknown** | 5 | 5 | 6 | **4** |

given that the algorithm was trained on 3 colours and 3 shapes, we would expect that it should recognize different hues of the 3 colours and objects with similar shapes to the original 3; however, it may not be able to recognize objects with completely different features. Moreover, we further group different symbols into concept groups. If any two symbols are described by the same features, it is safe to assume that those two symbols are mutually exclusive: that is, they can not both describe an object simultaneously. Thus we go over each concept group and if there are symbols yielding PDF values above a predefined threshold, we assign the new image the symbol from that group with the highest PDF.

## 3.3 Results

The system successfully learns from the training dataset that the colour symbols are being characterized by the extracted RGB values, while (in contrast) the shape symbols from the pixel area of the image patch—see Figure 7. Given a new test image with its extracted features, the algorithm recognises 93% of presented colours and 56% of presented shapes. Tables 1 and 2 report the confusion matrices for the testing set. This shows that the system is more robust when recognizing colours than when recognizing shapes. This can be attributed to the fact that while RGB values describe the concept of colour well enough, simply the pixel area is not enough to describe the concept of shape. Therefore the algorithm confuses the rubber duck with a cell, for example, and the arrow with a cube, see Figure 8, principally because they are of a similar size to each other! In future work, we would consider a wider range of features being extracted from the images, which in turn would support a finer-grained discrimination among objects.

## 4   Discussion and Future Work

The experiments in this paper have demonstrated that it is possible to train classifiers for object appearance alongside symbols, which are analysed via a semantic parser, to achieve grounding of instructions that respect the specificity of the scenario within which that association derives its meaning. Although our framework supports an entire pipeline, from raw cross-modal input to an interpreted and grounded instruction, the presented scenarios are simple and the specific methods could be made (much) more sophisticated. Despite this, we claim that this provides stepping stones towards learning more complex language structures in the future: during the first few demonstrations a human could teach a robot fundamental concepts like colours, shapes, orientation, and then proceed to use this newly gained knowledge to ground, e.g., prepositional phrases (Forbes et al., 2015; Rosman and Ramamoorthy, 2011) or actions (Misra et al., 2016; Zampogiannis et al., 2015) in an online and context specific manner. Once the system knows what blue cubes look like, it would be easier to learn what it means for another cube to be on top of/around it.

Another fruitful line of exploration would be continuous learning of both known and unknown symbols, using the same conceptual groups the system has extracted from the training data. For instance, whenever the robot observes a new object it can either label it with a symbol or deem it as unknown for a particular concept group. Whenever a symbol is assigned, the feature model for that symbol is updated, taking into account the new data point. If on the other hand the symbol

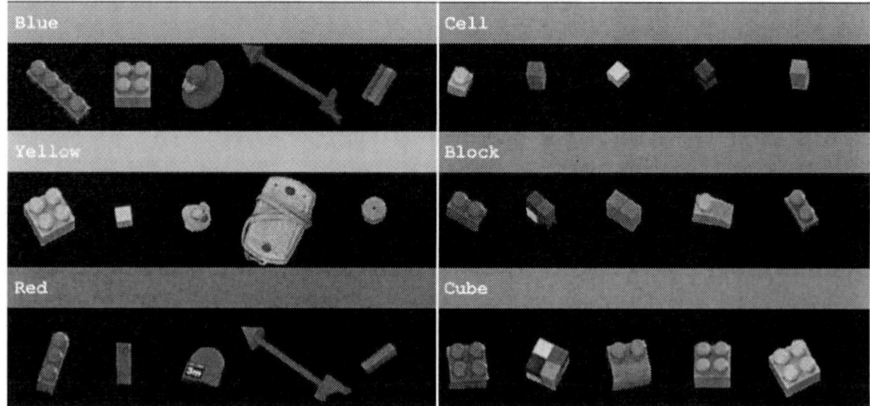


Figure 7: Excerpt from the testing dataset of objects whose colour and shape were correctly recognised.

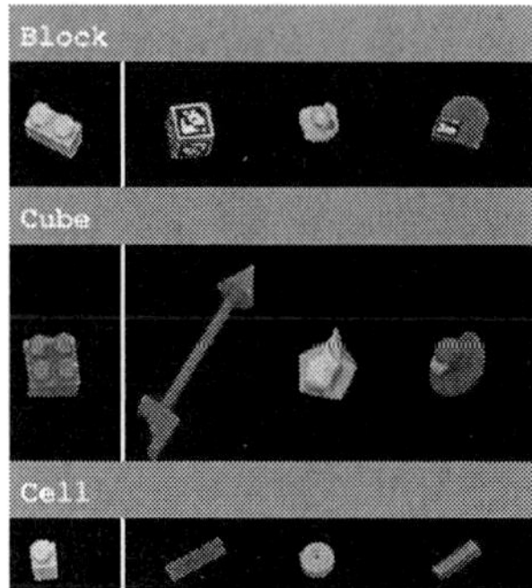


Figure 8: Examples of wrongly assigned known symbols to unseen shapes. Leftmost objects demonstrate an object from the training data.

is unknown, the system can prompt the human for new linguistic input which together with its feature model is added to the knowledge base and allows for its future recognition. For example, if the robot observes a new hue of blue it would update its parameters for `blue` to account for that; whereas if it observes a new colour (e.g. green) it would ask the human for the unknown symbol and would record it for future reference.

The idea of teaching the system about compound nouns is also a relevant challenge and a possible extension of this work: our current setup relies on noun phrases consisting of predicative `Adjs` and a `Noun` (e.g. blue cube), and so we know that the associated image patch $X$ satisfies both the adjective and the noun—i.e., $blue(X)$ and $cube(X)$ are both true. However, this would not apply to a compound noun like steak knife: we know that the associated image patch $X$ satisfies $knife(X)$ but does not satisfy $steak(X)$.

Refinements to our model would be necessary in order to represent more complex symbol relations, e.g. in a hierarchical fashion (Sun et al., 2014).

## 5 Conclusion

We present a framework for using cross-modal input: a combination of natural language instructions, video and eye tracking streams, to simultaneously perform semantic parsing and grounding of symbols used in that process within the physical environment. This is achieved without reliance on pre-existing object models, which may not be particularly representative of the specifics of a particular user's contextual usage and assignment of meaning within that rich multi-modal stream. Instead, we present an online approach that exploits the pragmatics of human sensorimotor behaviour to derive cues that enable the grounding of symbols to objects in the stream. Our preliminary experiments demonstrate the usefulness of this framework, showing how a robot is not only able to learn a human's notion of colour and shape, but also that it is able to generalise to the recognition of these features in previously unseen objects from a small number of physical demonstrations.

## Acknowledgments


This work is partly supported by ERC Grant 269427 (STAC), a Xerox University Affairs Committee grant, and grants EP/F500385/1 and BB/F529254/1 for the DTC in Neuroinformatics and Computational Neuroscience from the UK EPSRC, BBSRC, and the MRC. We are very grateful for the feedback from anonymous reviewers.

## References

M Al-Omari, P Duckworth, DC Hogg, and AG Cohn. 2016. Natural language acquisition and grounding for embodied robotic systems. In *Proceedings of the 31st Association for the Advancement of Artificial Intelligence*. AAAI Press.

Ann Copestake, Dan Flickinger, Carl Pollard, and Ivan A Sag. 2005. Minimal recursion semantics: An introduction. *Research on Language and Computation* 3(2-3):281–332.

Miles Eldon, David Whitney, and Stefanie Tellex. 2016. Interpreting multimodal referring expressions in real time. In *International Conference on Robotics and Automation*.

Dan Flickinger, Emily M. Bender, and Stephan Oepen. 2014. ERG semantic documentation. Accessed on 2017-04-29. http://www.delph-in.net/esd.

Maxwell Forbes, Rajesh PN Rao, Luke Zettlemoyer, and Maya Cakmak. 2015. Robot programming by demonstration with situated spatial language understanding. In *Robotics and Automation (ICRA), 2015 IEEE International Conference on*. IEEE, pages 2014–2020.

Thomas Kollar, Jayant Krishnamurthy, and Grant P Strimel. 2013. Toward interactive grounded language acqusition. In *Robotics: Science and Systems*.

Cynthia Matuszek, Liefeng Bo, Luke Zettlemoyer, and Dieter Fox. 2014. Learning from unscripted deictic gesture and language for human-robot interactions. In *AAAI*. pages 2556–2563.

Cynthia Matuszek, Evan Herbst, Luke Zettlemoyer, and Dieter Fox. 2013. Learning to parse natural language commands to a robot control system. In *Experimental Robotics*. Springer, pages 403–415.

Dipendra K Misra, Jaeyong Sung, Kevin Lee, and Ashutosh Saxena. 2016. Tell me dave: Context-sensitive grounding of natural language to manipulation instructions. *The International Journal of Robotics Research* 35(1-3):281–300.

Marcin Morzycki. 2013. Modification. Book manuscript. In preparation for the Cambridge University Press series *Key Topics in Semantics and Pragmatics*. http://msu.edu/ morzycki/work/book.

Raúl Mur-Artal, J. M. M. Montiel, and Juan D. Tardós. 2015. ORB-SLAM: a versatile and accurate monocular SLAM system. *IEEE Transactions on Robotics* 31(5):1147–1163. https://doi.org/10.1109/TRO.2015.2463671.

Jiquan Ngiam, Aditya Khosla, Mingyu Kim, Juhan Nam, Honglak Lee, and Andrew Y Ng. 2011. Multimodal deep learning. In *Proceedings of the 28th international conference on machine learning (ICML-11)*. pages 689–696.

Stephan Oepen, Dan Flickinger, Kristina Toutanova, and Christopher D Manning. 2004. *Research on Language and Computation* 2(4):575–596.

Natalie Parde, Adam Hair, Michalis Papakostas, Konstantinos Tsiakas, Maria Dagioglou, Vangelis Karkaletsis, and Rodney D Nielsen. 2015. Grounding the meaning of words through vision and interactive gameplay. In *IJCAI*. pages 1895–1901.

Svetlin Penkov, Alejandro Bordallo, and Subramanian Ramamoorthy. 2017. Physical symbol grounding and instance learning through demonstration and eye tracking .

Benjamin Rosman and Subramanian Ramamoorthy. 2011. Learning spatial relationships between objects. *The International Journal of Robotics Research* 30(11):1328–1342.

Constantin A Rothkopf, Dana H Ballard, and Mary M Hayhoe. 2007. Task and context determine where you look. *Journal of vision* 7.

Nitish Srivastava and Ruslan R Salakhutdinov. 2012. Multimodal learning with deep boltzmann machines. In *Advances in neural information processing systems*. pages 2222–2230.

Yuyin Sun, Liefeng Bo, and Dieter Fox. 2014. Learning to identify new objects. In *Robotics and Automation (ICRA), 2014 IEEE International Conference on*. IEEE, pages 3165–3172.

Jesse Thomason, Jivko Sinapov, Maxwell Svetlik, Peter Stone, and Raymond J Mooney. 2016. Learning multi-modal grounded linguistic semantics by playing i spy. In *Proceedings of the Twenty-Fifth international joint conference on Artificial Intelligence (IJCAI)*.

Josh Tobin, Rachel Fong, Alex Ray, Jonas Schneider, Wojciech Zaremba, and Pieter Abbeel. 2017. Domain randomization for transferring deep neural networks from simulation to the real world. *arXiv preprint arXiv:1703.06907* .

Paul Vogt. 2002. The physical symbol grounding problem. *Cognitive Systems Research* 3(3):429–457.

Matthew R Walter, Sachithra Hemachandra, Bianca Homberg, Stefanie Tellex, and Seth Teller. 2014. A framework for learning semantic maps from grounded natural language descriptions. *The International Journal of Robotics Research* 33(9):1167–1190.

Konstantinos Zampogiannis, Yezhou Yang, Cornelia Fermüller, and Yiannis Aloimonos. 2015. Learning the spatial semantics of manipulation actions through preposition grounding. In *Robotics and Automation (ICRA), 2015 IEEE International Conference on*. IEEE, pages 1389–1396.

# Exploring Variation of Natural Human Commands to a Robot in a Collaborative Navigation Task

Matthew Marge[1], Claire Bonial[1], Ashley Foots[1], Cory Hayes[1], Cassidy Henry[1],
Kimberly A. Pollard[1], Ron Artstein[2], Clare R. Voss[1], and David Traum[2]

[1]U.S. Army Research Laboratory, Adelphi, MD 20783
[2]USC Institute for Creative Technologies, Playa Vista, CA 90094
matthew.r.marge.civ@mail.mil

## Abstract

Robot-directed communication is variable, and may change based on human perception of robot capabilities. To collect training data for a dialogue system and to investigate possible communication changes over time, we developed a Wizard-of-Oz study that (a) simulates a robot's limited understanding, and (b) collects dialogues where human participants build a progressively better mental model of the robot's understanding. With ten participants, we collected ten hours of human-robot dialogue. We analyzed the structure of instructions that participants gave to a remote robot before it responded. Our findings show a general initial preference for including metric information (e.g., *move forward 3 feet*) over landmarks (e.g., *move to the desk*) in motion commands, but this decreased over time, suggesting changes in perception.

## 1 Introduction

Instruction-giving to robots varies based on perception of robots as conversational partners. We present an experiment designed to elicit robot-directed language that is a happy medium between existing natural language processing capabilities and fully natural communication. The data elicited will be used to train a dialogue system in the future, and it provides insights into what communication strategies people use when instructing robots. In this paper, we begin to examine how people vary their strategies as they build a progressively more accurate mental model of the robot and its capabilities. To simulate a robot's limited understanding of its environment, we employ the Wizard-of-Oz (WOz) method, where humans sim-

ulate robot intelligence and actions without participant awareness. With ten participants, we collected ten hours of human-robot dialogue. We are currently undertaking corpus curation and plan to make the data freely available in the next year.

In this experiment, a human and robot engage in a series of *transactions* (Carletta et al., 1997) where an instruction is issued, and wizards acting on behalf of the robot either perform a task or prompt for clarification until the requested task is completed or abandoned. We propose a new term, *instruction unit* (IU), to identify all commands within a transaction issued before the robot generates a response. IUs were analyzed both in structure and variation. Our findings suggest a general, initial preference for including metric information over landmarks in motion commands, but this decreased over time. Results will assist in future work adapting robot responses to varied instruction styles.

In the sections to follow, we first give needed background: the experiment setup and our approach to eliciting natural, robot-directed language. We then describe the annotations we have undertaken thus far to explore communication strategies. In our results section, we provide some statistics on the data collected thus far as well as noted changes in communication strategies. We provide a discussion of these results and comparison to related work and close with a summary and description of future work.

## 2 Background

### 2.1 Collaborative Exploration Task

The domain testbed for our work is collaborative exploration in a low-bandwidth environment (Marge et al., 2016). This testbed mimics what can be found in a reconnaissance or search-and-navigation operation, wherein a human *Com-*



*Proceedings of the First Workshop on Language Grounding for Robotics*, pages 58–66,
Vancouver, Canada, July 30 - August 4, 2017. ©2017 Association for Computational Linguistics

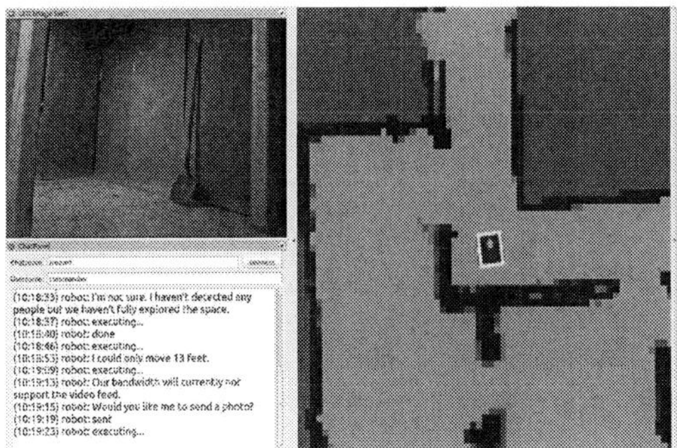

Figure 1: Interface showing robot text responses (*lower left*) to the Commander's verbal instructions, map (*right*), and the last still image sent (*upper left*).

*mander* verbally instructs a robot at a remote location, guiding the robot to move around and explore a physical space. The sensors and video camera on-board the robot populate a map as it moves, enabling it to describe that environment and send photos at the Commander's request, but the communications bandwidth prohibits real-time video streaming or direct teleoperation. The robot is assumed capable of performing low to intermediate level tasks, but not more complex tasks involving multiple or quantified goals, without clear directions or plans for ordering subgoals. The physical implementation of the testbed is an indoor environment, containing several rooms and connecting hallways, located in a separate building from the Commander. We use a Clearpath Robotics Jackal, fitted with an RGB camera and LIDAR sensors, to operate in the environment.

The Commander sees the following information from the robot's sensor data: a 2D occupancy grid with the robot's current position and heading streamed within the grid (i.e., map), and the last still image captured by its front-facing camera. In addition, the Commander can speak to the robot and see the robot's text responses. Figure 1 shows the information made available to the Commander.

## 2.2 Experiment Design

In each session, a (Commander) participant engaged the robot in collaborative search-and-navigation tasks. A session was comprised of three twenty-minute phases: a training phase and two main task phases (main phase 1 and 2). Training may voluntarily end when participants were comfortable with controls. Each phase focused

on a slightly different search task and started in a distinct location. Experiment tasks were developed to encourage the participant to use the robot as a teammate to search for certain objects in the environment. The participant needed to use their real-world knowledge in order to answer questions that required analysis of the observed environment. The robot didn't know common words for target objects, which required participants to consider word choice as they addressed the robot. An example search task was to locate *shoes* in an environment, relying on robot-provided images. An example analysis task was to consider whether the explored space was suitable as a headquarters-like environment. All phases situated the robot in an unfamiliar indoor environment, unlike canonical scenes typically observed in homes and offices.

Preceding the study, participants received a list of robot capabilities (see Appendix A). They were told that the robot understood basic object properties (e.g., most object labels, color, size), relative proximity, some spatial terms, and location history. Participants were not given example instructions.

## 2.3 Wizard-of-Oz Setup

We use a WOz approach to allow for understanding of natural domain-specific instructions, in advance of collecting enough training data to implement an automated system. Our work expands on existing WOz approaches by incorporating multimodal communication when the robot and human are not co-present – where information exchange of robot position, visual media, and dialogue is needed for collaborative exploration to succeed.

We use a multi-wizard setup to simulate the expected autonomous robot understanding and response. We use two wizards simultaneously for two reasons. First, a single wizard cannot type dialogue responses while teleoperating the robot with a joystick at the same time. Second, by design, we wish to decouple navigation behavior from dialogue behavior, as these will ultimately be separate modules in a fully-automated system.

A *Dialogue Manager* (DM-Wizard) listens to Commander speech and communicates directly with the Commander, using a chat window to type status updates and requests for clarification. When the Commander's instructions are executable in the current context, the DM-Wizard types in another chat window to pass a constrained, text in-

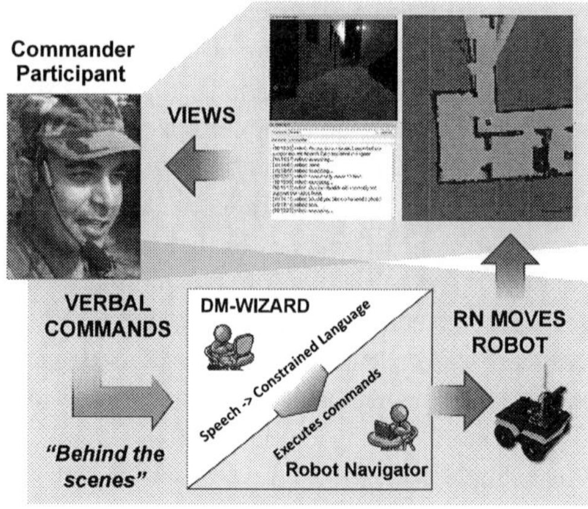


Figure 2: Wizard-of-Oz setup with wizards for dialogue management and robot navigation.

struction set to the *Robot Navigator* (RN), who teleoperates the robot. When hearing robot status updates directly from the RN, the DM-Wizard also communicates this information back to the participant. The DM-Wizard and RN roles were kept constant by having the same experimenters (female DM-Wizard, male RN) in those roles for the entirety of the study. Figure 2 presents our setup.

## 3 Approach: Eliciting Natural Language

One of the main research questions we seek to address with this experimental design is how to elicit natural communication, given that people may change strategies over time as they accommodate the robot's limited understanding. Like Chai et al. (2014) and Williams et al. (2015), we are interested in methods that robots can use to interpret and convey common ground in natural language interaction. Here, we describe how our DM-Wizard command-handling guidelines simulate a robot's limited understanding and the strategies that it could use to disambiguate phrases. Next, we introduce transaction and instruction units as a way to identify and measure possible variation in participant instructions.

### 3.1 DM-Wizard Guidelines

One way to elicit natural communications is to have the robot (in this case, the DM-Wizard) use strategies that mitigate its limited understanding, like offering suggestions or conveying its capabilities. We developed guidelines to determine when to employ such strategies and to ensure consistent dialogue decisions across participants. The

**Participant command** (speech): *Move forward.*
**Communication problem**: Open-ended action (no endpoint specified)
**Relevant template**:
DESCRIBE PROBLEM + CAPABILITY
**DM-Wizard response to participant** (text): `How far? You can tell me to move to an object that you see or a distance.`
**Participant response** (speech): *Move to the yellow cone ahead of you.*

Figure 3: DM-Wizard guidelines for consistent dialogue behaviors (developed iteratively in piloting) applied in a sample exchange.

guidelines governed the DM-Wizard's real-time decision-making. They first identify the minimal requirements for an executable command: each must contain both a clear action and respective endpoint. The guidelines provide response categories and templates, allowing for flexibility in exact response form, but with easily-remembered templates for elements of each response. Responses are broadly categorized into well-formed vs. unclear, problematic commands. The exchange in Figure 3 shows how a participant's problematic, open-ended instruction is handled under the guidelines.

### 3.2 Dialogue Structure Annotation

In order to both study the question of what kinds of language, discourse, and dialogue strategies are used to give instructions, as well as to provide training data for automating the DM-Wizard functions, we annotated several aspects of dialogue structure. In this paper we focus on the former question, and examine how participants convey initial task intention to a robot before follow-on dialogue from the DM-Wizard. This analysis helps us understand the structure of instructions and anticipate possible task ranges required of a robot. We discuss four levels of dialogue semantics and structure below, from largest to smallest: *transaction units* (TUs), *instruction units* (IUs), *dialogue-moves*, and *parameters*. Each of these is defined and discussed below.

### 3.2.1 Transaction Units

Each dialogue is annotated as a series of higher-level *transaction units*. A TU is a sequence of utterances aiming to achieve task intention. The TUs document structures that appear within collected dialogues, while also providing emulatable interaction patterns for a dialogue manager. TUs

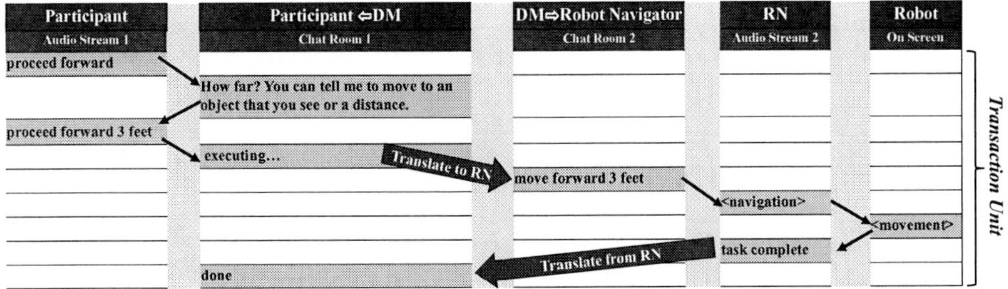


Figure 4: Two wizards manage the labor of robot intelligence. Dialogues divide into a series of *transactions* where a naive participant gives an instruction, a *Dialogue Manager* (DM-Wizard) decides how to handle it, and passes well-formed instructions to a *Robot Navigator* (RN) that moves the robot.

each contain a participant's initiating message and then subsequent messages by the participant and wizards to complete the transaction.

Figure 4 shows an example transaction in which a participant gives instructions, the DM-Wizard requests clarification, and the amended instructions are then passed to the RN, who completes the instructions.

### 3.2.2 Instruction Units

Within TUs, we marked *instruction units*. An IU comprises all participant speech to the robot within a transaction unit before robot feedback. Each IU belongs to exactly one TU, so that the start of each transaction (e.g., a new command is issued) marks a new IU. An IU terminates when the robot replies to the request, or when a new transaction is initiated. The relationships of IUs and TUs is shown in Figure 5.

### 3.2.3 Dialogue-Moves

To analyze internal IU structure, we annotated Commander-issued lower-level *dialogue-moves*. This annotation scheme is inspired by a prior approach to military dialogue that identified dialogue-moves in calls for artillery fire (Roque et al., 2006). Examples of a *command* type request are *command:drive* or *command:rotate*, that instruct the robot to perform certain motions. A dialogue-move list is provided in Appendix B.

Three annotators independently validated the dialogue-move set on 99 dialogue turns in our human-robot dialogue corpus. Annotators had high agreement ($\alpha = 0.92$; Krippendorf's $\alpha$ using the MASI distance measure (Passonneau, 2006)).

### 3.2.4 Parameters on Motion Commands

Some dialogue moves uniquely define the action that the robot should take, e.g. *command:stop* or

*command:send-image*. Others require additional *parameters* to fully specify the complete action. Of particular interest to us is the information that participants chose to include in robot-directed motion requests. We focused on *command:drive* and *command:rotate* for variation in how participants communicated. We annotated motion-command parameters for their usage of *metric* (e.g., *move forward 2 feet; turn left 90 degrees*) and *landmark*-based points of reference (e.g., *move to the table; turn to face the doorway*) similar to *absolute* and *relative* steps in route instructions from Marge and Rudnicky (2010).

### 3.3 Participants

This study recruited ten participants: two female, eight male. Ages ranged from 28 to 58 (mean = 44, s.d. = 10.6). Two participants reported one year or less of robotics research; others reported none.[1]

Figure 5: Annotation structures on human-robot dialogue, shown over participant and DM-Wizard streams.

---

[1] We collected measures that were included in our statistical analysis but not presented in this paper. The Spatial Orientation Survey, part of the Guilford-Zimmerman Aptitude Survey (Guilford and Zimmerman, 1948), assesses spatial orientation perception. The HRI Trust Survey (Schaefer, 2013) measures subjective trust of the robot, based upon personal belief of the robot's capabilities.

### 3.4 Corpus Statistics

We collected approximately 10.5 hours of recorded Commander speech (approximately 1 hour per participant), and DM-Wizard text messages to participants and to the RN. All live video feed, map, and robot pose data, as well as task-relevant images requested by participants, were recorded. Language data was manually time-aligned. After transcription and annotation, the corpus yielded 858 IUs.

## 4 Results

Each IU in the corpus corresponded to a unique TU from participant-robot dialogue. To better understand the structure and possible instruction variation over time, we focused analysis on IUs, their respective dialogue-moves, and motion command parameters. We analyzed IUs based on measures of word count, dialogue-move, and parameters on motion commands. We assessed possible parametric differences on motion commands by experiment phase (training phase, main phase 1, main phase 2). For significance testing, we used a mixed-effects ANOVA (computing standard least square regression using reduced maximum likelihood (Harville, 1977)), where phase (a repeated measure), age, gender, and scores on the spatial orientation and HRI trust surveys were included as factors in the model. Participant ID was included as a random effect.

### 4.1 Instruction Units

To gauge instruction frequency, we observed the mean number of IUs issued in an experiment session. On average, each participant issued 86 IUs (s.d. = 24.7, min = 58, max = 126). The average IU length was 8 words (s.d. = 5.7, min = 1, max = 60). Three participants each issued over 112 IUs in total, while three issued 70 or fewer.

### 4.2 Dialogue-Moves in IUs

We analyzed the selection of dialogue-moves that participants issued in their IUs. Participants often issued more than one dialogue-move per IU (mean = 1.6 dialogue-moves per IU, s.d. = 0.88, min = 1, max = 8). Unsurprisingly, the *command* dialogue-move was in the most IUs (94% of all IUs). See Table 1 for the entire distribution. We report on notable exceptions in Section 5.2.

The most common functions observed in the instructions were *command* dialogue-moves to send

| Dialogue-Move | Instruction Units | |
|---|---|---|
| | N | % |
| *Command* | | 94 |
|   *Send-Image* | 443 | 52 |
|   *Rotate* | 406 | 47 |
|   *Drive* | 358 | 42 |
|   *Stop* | 29 | 3 |
|   *Explore* | 7 | 1 |
| *Request-Info* | 34 | 4 |
| *Feedback* | 28 | 3 |
| *Parameter* | 14 | 2 |
| *Describe* | 5 | 1 |

Table 1: Dialogue-move distribution over all IUs in the corpus (N=858). An IU may have one or more dialogue-moves.

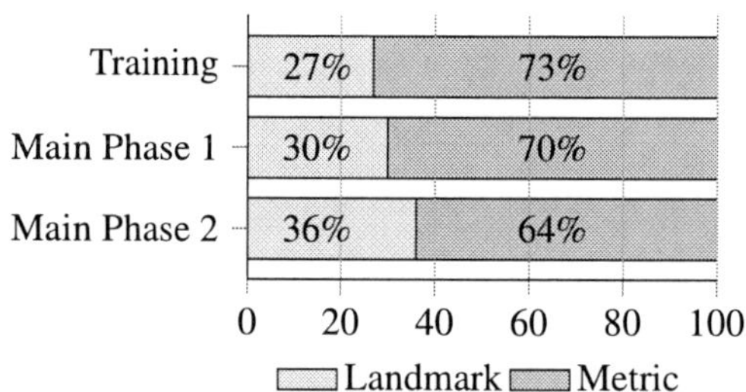


Figure 6: Proportions of landmark mentions to metric mentions within all *command* moves of subtype *drive* and *rotate* across experiment phases. There were 177, 333, and 316 occurrences of metric or landmark information in the training, main phase 1, and main phase 2 to compute proportions, respectively.

a new image, rotate, and drive. As reported in Table 1, over half of IUs include an image request, followed by rotate and drive commands.

### 4.3 Parameters on Motion Commands

We delineate percentages of all IUs that involved motion requests for the robot (i.e., commands that were not image, stop, or exploration requests). 638 IUs contained a *drive* or *rotate* subtype request with a command parameter; 75% included metric units and 37% included landmarks (an IU could contain both). We tabulated all metric and landmark mentions in this IU subset.

We observed a substantial change in general participant strategy over time (Figure 6). In the training phase, participants began with a metric-dominant strategy that regressed in main phase 1, and further in main phase 2. The final phase experienced a 9% increase (absolute) in landmark references compared to the training phase, and a subsequent 9% decrease of metric references. A mixed-effects ANOVA test on the proportion of metric to landmark usage in commands found a main effect for phase (F[2, 627]=3.6, p<0.05). No other main effects were found. A Tukey HSD test found a significant difference between main phase 1 and 2 (p<0.05). We also tabulated instances of increased landmark usage by participant: six participants increased their proportion of landmark usage between main phase 1 and 2. Three used fewer landmarks in main phase 2, and one used the same proportion.

## 5  Discussion

This work seeks to elicit natural instruction-giving from participants, and to assess how communication strategies varied as people build an increasingly better mental model of a robot's understanding. Thus far, we've seen progress towards our goal in two main areas: (1) the experiment setup was workable; participants believed they were instructing an autonomous robot, and (2) we observed naturally occurring coordination efforts via changes in participant strategy over time. The latter area is discussed in more detail in the next subsection.

### 5.1  Metric vs. Landmark Usage

Our findings suggest possible changes in how participants perceived robot capabilities over time. This was highlighted by a significant decrease in metric usage between the two main experiment phases (main phase 1 and 2). This result suggests that participants became more comfortable in communicating with the robot through experience. Therefore, their communication styles become more "natural" and similar to human communication strategies, which tend to include landmark-based references (Clarke et al., 2015). This result has implications for language grounding and interpretation, in that developers should expect to handle both metric and landmark-based references.

We note that the dominant strategy overall was clearly the use of metric information. We identify several possible factors. One factor may be the participant interface: their situational awareness of the robot's environment is constrained to the most recent image of the robot's first-person perspective of a scene and the map displaying an occupancy grid of the surroundings. The indoor space is sparsely populated with objects, so a requested image might not return valuable visual information of an object of interest. The map, on the other hand, is visually salient and returns real-time information, including the presence of rooms, halls, and doorways. When landmark references are combined, the most frequent landmark used (139 out of 380 total landmark references) is "door," followed by "room," "hallway," and "wall." These are all landmarks recognizable in the map; other landmark types for navigation may be inhibited due to participant unawareness.

A second, somewhat related factor contributing to the use of metric references in general may be a misalignment of common ground between participant and robot, namely a lack of familiarity with the objects. Even when an object is returned in an image, the angle may not be conducive to object recognition. Participants are forced to either abandon the object as a landmark, or find another way of talking about it. For example, one participant describes a calendar hanging on the wall as *"the item on the right on the wall,"* while another describes a barrel as *"the round object."*

In addition, participants may be unsure if the robot can recognize an object by a given word. In training, participants were instructed that "the robot knows what some objects are but not all objects." They also know that the robot understands object features. Our intention was to encourage dialogue by making high-level search commands like *"find the shoe"* (a search-task target) outside the robot's capabilities. A side effect is that participants quickly became aware of this limitation, often as early as the training phase. When participants did try including a search-task target, without any additional descriptive information like color or shape, the DM-Wizard guidelines prompted for an alternative description. Some participants ignored the robot's request for a different description, and instead abandoned the landmark strategy in favor of metric instructions, which can be used in the absence of familiarity or knowledge of surrounding landmarks.

We note that the robot's surroundings are somewhat strange. They do not conform to canonical representations, disallowing use of lived experience of object expectations based on room type. Although an effort was made to group similar objects according to a room's possible function (e.g., kitchen items grouped together in one room and in a typical arrangement), the environment is sparsely filled with objects and is not in a finished state. These were practical limitations of laboratory resources, but in future work we plan to explore the effects of the environment further by varying it in a fully simulated version of the experiment.

## 5.2 Dialogue-Move Types

We found that most IUs contained *command* dialogue-moves, but with some exceptions. This was largely based on participants' assessment of robot capabilities. Two participants were responsible for 33 of the 34 occurrences of *request-info*. One participant issued requests like *"are you alone?"* and *"do you detect any threats?"* The other requested object identification, such as *"what's that object just to the left of the photo?"* This suggests an expectation for additional joint vision and language processing capabilities in these kinds of scenarios. *Feedback* dialogue-moves were largely experiment-specific start and end updates like *"I am ready."*

Our dialogue-move analysis of *commands* revealed a uniform strategy of consistent image requests shown in nearly half of all IUs. This is expected, as the bandwidth limitations of our experiment design prevented sending live video. More image requests are expected, but we found at least five phase runs where the robot "learned" to send images after receiving commands: occasionally the DM-Wizard would observe that a participant was requesting an image in every instruction, and as a result offered to remember to send images after each command.

## 6 Related Work

Our experiment setup and data collection effort resemble similar corpora, with some differences. The CReST (Eberhard et al., 2010), SCARE (Stoia et al., 2008), and GIVE (Gargett et al., 2010) corpora consist of search-and-navigation tasks, but are strictly human-human dialogue. We collected natural language interactions simulating fully autonomous dialogue processing, but without participant awareness that a human was simulating the robot responses. Participants assessed robot in-

telligence on their own when formulating instructions and follow-on responses.

## 6.1 Wizard-of-Oz Approach

By far the WOz method's most common use has been for handling natural language (Riek, 2012). Many studies use a wizard in automated dialogue system development (e.g., in virtual agent negotiation (Gandhe and Traum, 2007), time-offset storytelling (Artstein et al., 2015), and in-car personal assistants (Rieser and Lemon, 2008)).

Some researchers have considered a multi-wizard setup for multimodal interfaces. The Sim-Sensei project (DeVault et al., 2014) used a two-wizard setup during the development stage; one controlling the virtual agent's verbal behaviors and another the non-verbal behaviors. Green et al. (2004) investigated using multiple wizards for dialogue processing and navigation capabilities for a robot in a home touring scenario, finding the multi-wizard approach effective when the robot and human were co-present.

## 6.2 Natural Language Interpretation

Traditional approaches to natural language interpretation for robots follow the methodology of *corpus-based robotics* (Bugmann et al., 2004), where some natural language, primarily route instructions, is collected. Route instruction interpreters dating back to MARCO (MacMahon, 2006), and more recently the robotic forklift (Tellex et al., 2011) and Tactical Behavior Specification grammar (Hemachandra et al., 2015; Boularias et al., 2016), rely on these initial route instructions to learn mappings to robot-executable procedures like path planning. Additionally, some use semantic parsers (e.g., (Chen and Mooney, 2011; Artzi and Zettlemoyer, 2013; Matuszek et al., 2013; Krishnamurthy and Kollar, 2013)) or translation (Matuszek et al., 2010) to map natural language to actions.

A gap in these works is bi-directional dialogue interaction, specifically cases where initial instructions are not well-formed and need additional clarification, or when participants grow to better grasp the robot's capabilities, varying instruction strategies over time. Our work collected instructions to a robot, but also included the dialogue and follow-on responses needed to establish or build common ground. This paper focused on analyzing initial robot-directed instructions, leaving analysis of responses during the dialogue to future work.

## 7 Summary and Ongoing Work

We presented a method for investigating changes in participant instruction strategies to a robot in a collaborative navigation task. We found an initial preference for metric information in motion commands, but this decreased over time as participants used more landmarks in their instructions.

We also note that the dataset under construction will provide value not only in the language collected, but also visual information. The accompanying images from the robot provide a unique resource with content that is both first-person and task-relevant for building situational awareness of a remote environment.

This work is a multi-stage effort to develop natural communication frameworks between humans and robots. In this work's next phase, automating language processing will begin, starting with language generation aspects. Rather than typing out full responses, wizards will use an interface to select responses following communicative guidelines. The Wizard-of-Oz interface allows template generation by filling in parameter values, if necessary. We expect a similar range of rich participant dialogue, but faster wizard response time, even for fairly complex strategies. Wizard selections will serve as training data for an automated dialogue manager.

## Acknowledgments

We thank Susan G. Hill, Arthur William Evans, Stephanie Lukin, Su Lei, and Brendan Byrne for their contributions to this research program.

## Appendix

## A Robot Capabilities

These are, verbatim, the capabilities provided on a sheet to study participants:

"The robot can take a photo of what it sees when you ask. The robot has certain capabilities, but cannot perform these tasks on its own. The robot and you will act as a team.
Robot capabilities are:

- Robot listens to verbal instructions from you.

- Robot responds in this text box *(Experimenter points to instant messenger box on screen)* or by taking action

- Robot will avoid obstacles

- Robot can take photos directly in front of it when you give it a verbal instruction

- Robot will know what some objects are, but not all objects

- Robot also knows:
  - Intrinsic properties like color and size of objects in the environment
  - Proximity of objects like where objects are relative to itself and to other objects
  - A range of spatial terms like to the right of, in front of, cardinal directions like N, S
  - *History:* the Robot remembers places it has been

- Robot doesn't have arms and it cannot manipulate objects or interact with its environment except for moving throughout the environment

- Robot cannot go through closed doors and it cannot open doors, but it can go through doorways that are already open

- Robot can only see about knee height ($\sim$ 1.5 feet)."

## B Dialogue-Move Annotation Set

**Command** Task-related instructions from the Commander to the robot are *command* dialogue-moves.

- *command:drive* Initiate/continue movement.
- *command:rotate* Initiate/continue a rotation.
- *command:explore* Explore an area via navigation using a target and/or direction as heading.
- *command:stop* End a drive or rotation.
- *command:send-image* Request an image.

**Describe** General statements from the Commander to the robot about a scene or plan are *describe* dialogue-moves.

- *describe:scene* Typically a description of what the Commander sees or thinks the robot should see.
- *describe:plan* Explication of the Commander's intention, not necessarily actionable.

**Request-info** *Request-info* dialogue-moves request information of the robot.

- *request-info:scene* Asking for information about what the robot sees, or confirmation for what the Commander thinks the robot should see.
- *request-info:map* Asking about robot's position or heading.
- *request-info:confirm* Confirm a proposed plan.

**Feedback** General domain-independent expressions from the Commander to the robot.

- *acknowledge* Acknowledgment of either a conversational move or an action (such as the sending of an image or map).
- *ready* Inform robot ready to do task.
- *yes* Simple positive response (yes).
- *no* Simple negative response (no).
- *standby* Inform robot to stand by or wait.

**Standalone Instruction Content** Provide further content for an existing instruction from the Commander to the robot.

- *direction* a heading (e.g., right, left)
- *distance* a unit of measure (e.g., feet, degrees)

## References

Ron Artstein, Anton Leuski, Heather Maio, Tomer Mor-Barak, Carla Gordon, and David R Traum. 2015. How many utterances are needed to support time-offset interaction? In *Proc. of FLAIRS*.

Yoav Artzi and Luke Zettlemoyer. 2013. Weakly supervised learning of semantic parsers for mapping instructions to actions. *Transactions of the Association for Computational Linguistics* 1:49–62.

Abdeslam Boularias, Felix Duvallet, Jean Oh, and Anthony Stentz. 2016. Learning qualitative spatial relations for robotic navigation. In *Proceedings of IJCAI*.

Guido Bugmann, Ewan Klein, Stanislao Lauria, and Theocharis Kyriacou. 2004. Corpus-Based Robotics: A Route Instruction Example. In *Proc. of IAS-8*.

Jean Carletta, Stephen Isard, Gwyneth Doherty-Sneddon, Amy Isard, Jacqueline C Kowtko, and Anne H Anderson. 1997. The reliability of a dialogue structure coding scheme. *Computational linguistics* 23(1):13–31.

Joyce Y Chai, Lanbo She, Rui Fang, Spencer Ottarson, Cody Littley, Changsong Liu, and Kenneth Hanson. 2014. Collaborative effort towards common ground in situated human-robot dialogue. In *Proc. of HRI*.

David L Chen and Raymond J Mooney. 2011. Learning to Interpret Natural Language Navigation Instructions from Observations. In *Proc. of AAAI*.

Alasdair DF Clarke, Micha Elsner, and Hannah Rohde. 2015. Giving good directions: order of mention reflects visual salience. *Frontiers in psychology* 6:1793.

David DeVault, Ron Artstein, Grace Benn, Teresa Dey, Ed Fast, Alesia Gainer, Kallirroi Georgila, Jon Gratch, Arno Hartholt, Margaux Lhommet, et al. 2014. SimSensei Kiosk: A virtual human interviewer for healthcare decision support. In *Proc. of AAMAS*.

Kathleen M. Eberhard, Hannele Nicholson, Sandra Kübler, Susan Gundersen, and Matthias Scheutz. 2010. The Indiana "Cooperative Remote Search Task" (CReST) Corpus. In *Proc. of LREC*.

Sudeep Gandhe and David R Traum. 2007. Creating spoken dialogue characters from corpora without annotations. In *Proc. of Interspeech*.

Andrew Gargett, Konstantina Garoufi, Alexander Koller, and Kristina Striegnitz. 2010. The GIVE-2 Corpus of Giving Instructions in Virtual Environments. In *Proc. of LREC*.

Anders Green, Helge Huttenrauch, and Kerstin Severinson Eklundh. 2004. Applying the Wizard-of-Oz framework to cooperative service discovery and configuration. In *Proc. of ROMAN*.

Joy Paul Guilford and Wayne S Zimmerman. 1948. The Guilford-Zimmerman Aptitude Survey. *Journal of applied Psychology* 32(1):24.

David A. Harville. 1977. Maximum likelihood approaches to variance component estimation and to related problems. *Journal of the American Statistical Association* 72(358):320–338.

Sachithra Hemachandra, Felix Duvallet, Thomas M Howard, Nicholas Roy, Anthony Stentz, and Matthew R Walter. 2015. Learning models for following natural language directions in unknown environments. In *Proc. of ICRA*.

Jayant Krishnamurthy and Thomas Kollar. 2013. Jointly Learning to Parse and Perceive: Connecting Natural Language to the Physical World. *Transactions of the Association for Computational Linguistics* 1:193–206.

Matt MacMahon. 2006. Walk the Talk: Connecting language, knowledge, and action in route instructions. In *Proc. of AAAI*.

Matthew Marge, Claire Bonial, Brendan Byrne, Taylor Cassidy, A. William Evans, Susan G. Hill, and Clare Voss. 2016. Applying the Wizard-of-Oz Technique to Multimodal Human-Robot Dialogue. In *Proc. of RO-MAN*.

Matthew Marge and Alexander I. Rudnicky. 2010. Comparing spoken language route instructions for robots across environment representations. In *Proc. of SIGdial*.

Cynthia Matuszek, Dieter Fox, and Karl Koscher. 2010. Following directions using statistical machine translation. In *Proc. of HRI*.

Cynthia Matuszek, Evan Herbst, Luke Zettlemoyer, and Dieter Fox. 2013. Learning to parse natural language commands to a robot control system. In *Experimental Robotics*.

Rebecca Passonneau. 2006. Measuring Agreement on Set-valued Items (MASI) for Semantic and Pragmatic Annotation. In *Proc. of LREC*.

Laurel Riek. 2012. Wizard of Oz Studies in HRI: A Systematic Review and New Reporting Guidelines. *Journal of Human-Robot Interaction* 1(1).

Verena Rieser and Oliver Lemon. 2008. Learning Effective Multimodal Dialogue Strategies from Wizard-of-Oz Data: Bootstrapping and Evaluation. In *Proc. of ACL*.

Antonio Roque, Anton Leuski, Vivek Rangarajan, Susan Robinson, Ashish Vaswani, Shrikanth Narayanan, and David Traum. 2006. Radiobot-CFF: A Spoken Dialogue System for Military Training. In *Proc. of Interspeech*.

Kristin E Schaefer. 2013. *The perception and measurement of human-robot trust*. Ph.D. thesis, University of Central Florida.

Laura Stoia, Darla Magdalena Shockley, Donna K. Byron, and Eric Fosler-Lussier. 2008. SCARE: A Situated Corpus with Annotated Referring Expressions. In *Proc. of LREC*.

Stefanie A. Tellex, Thomas F. Kollar, Steven R. Dickerson, Matthew R. Walter, Ashis Banerjee, Seth Teller, and Nicholas Roy. 2011. Understanding natural language commands for robotic navigation and mobile manipulation. In *Proc. of AAAI*.

Tom Williams, Gordon Briggs, Bradley Oosterveld, and Matthias Scheutz. 2015. Going Beyond Literal Command-Based Instructions: Extending Robotic Natural Language Interaction Capabilities. In *Proc. of AAAI*.

# A Tale of Two DRAGGNs:
# A Hybrid Approach for Interpreting
# Action-Oriented and Goal-Oriented Instructions


**Siddharth Karamcheti, Edward C. Williams, Dilip Arumugam,**
**Mina Rhee, Nakul Gopalan, Lawson L.S. Wong, Stefanie Tellex**
Department of Computer Science, Brown University, Providence, RI 02912
{siddharth_karamcheti@, edward_c_williams@, dilip_arumugam@,
mina_rhee@, ngopalan@cs., lsw@, stefie10@cs.}brown.edu


## Abstract


Robots operating alongside humans in diverse, stochastic environments must be able to accurately interpret natural language commands. These instructions often fall into one of two categories: those that specify a goal condition or target state, and those that specify explicit actions, or how to perform a given task. Recent approaches have used reward functions as a semantic representation of goal-based commands, which allows for the use of a state-of-the-art planner to find a policy for the given task. However, these reward functions cannot be directly used to represent action-oriented commands. We introduce a new hybrid approach, the Deep Recurrent Action-Goal Grounding Network (DRAGGN), for task grounding and execution that handles natural language from either category as input, and generalizes to unseen environments. Our robot-simulation results demonstrate that a system successfully interpreting both goal-oriented and action-oriented task specifications brings us closer to robust natural language understanding for human-robot interaction.


## 1 Introduction

Natural language affords a convenient choice for delivering instructions to robots, as it offers flexibility, familiarity, and does not require users to have knowledge of low-level programming. In the context of grounding natural language instructions to tasks, human-robot instructions can be interpreted as either high-level goal specifications or low-level instructions for the robot to execute.

Figure 1: Sample configuration of the Cleanup World mobile-manipulator domain (MacGlashan et al., 2015), used throughout this work. A possible goal-based instruction could be "Take the chair to the green room," while a possible action-based instruction could be "Go three steps south, then two steps west."

Goal-oriented commands define a particular target state specifying where a robot should end up, whereas action-oriented commands specify a particular sequence of actions to be executed. For example, a human instructing a robot to "go to the kitchen" outlines a goal condition to check if the robot is in the kitchen. Alternatively, a human providing the command "take three steps to the left" defines a trajectory for the robot to execute. We need to consider both forms of commands to understand the full space of natural language that humans may use to communicate their intent to robots. While humans also combine commands of both types into a single instruction, we make the simplifying assumption that a command belongs entirely to a single type and leave the task of handling mixtures and compositions to future work.

Existing approaches can be broadly divided into one of two regimes. Goal-based approaches like




*Proceedings of the First Workshop on Language Grounding for Robotics*, pages 67–75,
Vancouver, Canada, July 30 - August 4, 2017. ©2017 Association for Computational Linguistics

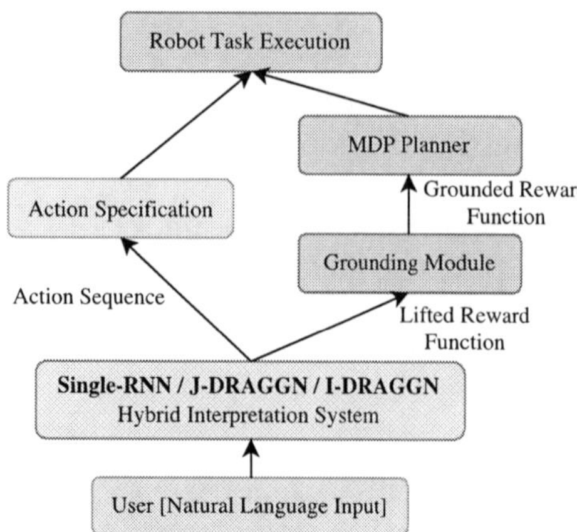

Figure 2: System for grounding both action-oriented (left branch) and goal-oriented (right branch) natural language instructions to executable robot tasks. Our main contribution is the hybrid interpretation system (blue box), for which we present two novel models based on the DRAGGN framework (J-DRAGGN and I-DRAGGN) in Section 4.

MacGlashan et al. (2015) and Arumugam et al. (2017) leverage some intermediate task representation and then automatically find a low-level trajectory to achieve the goal using a planner. Other approaches, in the action-oriented regime, directly infer action sequences (Tellex et al., 2011; Matuszek et al., 2012; Artzi and Zettlemoyer, 2013; Andreas and Klein, 2015) from the syntactic or semantic parse structure of natural language. However, these approaches can be computationally intractable for large state-action spaces or use ad-hoc methods to execute high-level language rather than relying on a planner. Furthermore, these methods are unable to adapt to dynamic changes in the environment; for example, consider an environment in which the wind, or some other force moves an object that a robot has been tasked with picking. Action sequence based approaches would fail to handle this without additional user input, while goal-based approaches would be able to re-plan on the fly, and complete the task.

To address the issue of dealing with both goal-oriented and action-oriented commands, we present a new language grounding framework that, given a natural language command, is capable of inferring the latent command type. Recent approaches leveraging deep neural networks have formulated the language grounding problem as sequence-to-sequence learning or multi-label classification (Mei et al., 2016; Arumugam et al., 2017). Inspired by the recent success of neural networks to model programs that are highly compositional and sequential in nature, we present the Deep Recurrent Action/Goal Grounding Network (DRAGGN) framework, derived from the the Neural Programmer-Interpreter (NPI) of Reed and de Freitas (2016) and outlined in Section 4.2. We introduce two instances of DRAGGN models, each with slightly different architectures. The first, the Joint-DRAGGN (J-DRAGGN) is defined in Section 4.3, while the second, the Independent-DRAGGN (I-DRAGGN) is defined in Section 4.4.

## 2 Related Work

There has been a broad and diverse set of work examining how best to interpret and execute natural language instructions on a robot platform (Vogel and Jurafsky, 2010; Tellex et al., 2011; Artzi and Zettlemoyer, 2013; Howard et al., 2014; Andreas and Klein, 2015; Hemachandra et al., 2015; MacGlashan et al., 2015; Paul et al., 2016; Mei et al., 2016; Arumugam et al., 2017). Vogel and Jurafsky (2010) produce policies using language and expert trajectories based rewards, which allow for planning within a stochastic environment along with re-planning in case of failure. (Tellex et al., 2011) instead grounds language to trajectories satisfying the language specification. (Howard et al., 2014) chose to ground language to constraints given to an external planner, which is a much smaller space to perform inference over than trajectories. MacGlashan et al. (2015) formulate language grounding as a machine translation problem, treating propositional logic functions as both a machine language and reward function. Reward functions or cost functions can allow richer descriptions of trajectories than plain constraints, as they can describe preferential paths. Additionally, Arumugam et al. (2017) simplify the problem from one of machine translation to multi-class classification, learning a deep neural network to map arbitrary natural language instructions to the corresponding reward function.

Informing our distinction between action sequences and goal state representation is the division presented by Dzifcak et al. (2009), who posited that natural language can be interpreted as *both* a goal state specification and an action specification. Rather than producing both from each language command, our DRAGGN framework makes the simplifying assumption that only one representation captures the semantics of the language; additionally, our framework does not require a manually pre-specified grammar.

Recently, deep neural networks have found widespread success and application to a wide array of problems dealing with natural language (Bengio et al., 2000; Mikolov et al., 2010, 2011; Cho et al., 2014; Chung et al., 2014; Iyyer et al., 2015). Unsurprisingly, there have been some initial steps taken towards applying neural networks to language grounding problems. Mei et al. (2016) uses a recurrent neural network (RNN) with long short-term memory (LSTM) cells (Hochreiter and Schmidhuber, 1997) to learn sequence-to-sequence mappings between natural language and robot actions. This model augments the standard sequence-to-sequence architecture by learning parameters that represent latent alignments between natural language tokens and robot actions. Arumugam et al. (2017) used an RNN-based model to produce grounded reward functions at multiple levels of an Abstract Markov Decision Process hierarchy (Gopalan et al., 2017), varying the abstraction level with the level of abstraction used in natural language.

Our DRAGGN framework is closely related to the Neural Programmer-Interpreter (NPI) (Reed and de Freitas, 2016). The original NPI model is a controller trained via supervised learning to interpret and learn when to call specific programs/subprograms, which arguments to pass into the currently active program, and when to terminate execution of the current program. We draw a parallel between inferred NPI programs and our method of predicting either lifted reward functions or action trajectories.

## 3   Problem Setting

We consider the problem of mapping from natural language to robot actions within the context of Markov decision processes. A Markov decision process (MDP) is a five-tuple $\langle S, A, T, R, \gamma \rangle$ defining a state space $S$, action space $A$, state transition probabilities $T$, reward function $R$, and discount factor $\gamma$ (Bellman, 1957; Puterman, 1994). An MDP solver produces a policy that maps from states to actions in order to maximize the total expected discounted reward.

While reward functions are flexible and expressive enough for a wide variety of task specifications, they are a brittle choice for specifying an exact sequence of actions, as enumerating every possible action sequence as a reward function (i.e. a specific reward function for the sequence Up 3, Down 2) can quickly become intractable. This paper introduces models that can produce desired behavior by inferring either reward functions or primitive actions. We assume that all available actions $A$ and the full space of potential reward functions (*i.e.*, the full space of possible tasks) are known *a priori*. When a reward function is predicted by the model, an MDP planner is applied to derive the resultant policy (see system pipeline Figure 2).

We focus our evaluation of all models on the the Cleanup World mobile-manipulator domain (MacGlashan et al., 2015; Arumugam et al., 2017). The Cleanup World domain consists of an agent in a 2-D world with uniquely colored rooms and movable objects. A domain instance is shown in Figure 1. The domain itself is implemented as an object-oriented Markov decision process (OO-MDP) where states are denoted entirely by collections of objects, with each object having its own identifier, type, and set of attributes (Diuk et al., 2008). Domain objects include rooms and interactable objects (*e.g* a chair, basket, etc.) all of which have location and color attributes. Propositional logic functions can be used to identify relevant pieces of an OO-MDP state and their attributes; as in MacGlashan et al. (2015) and Arumugam et al. (2017), we treat these propositional functions as reward functions. In Figure 1, the goal-oriented command "take the chair to the green room" may be represented with the reward function blockInRoom block0 room1, where the blockInRoom propositional function checks if the location attribute of block0 is contained in room1.

## 4   Approach

We now outline the pipeline that converts natural language input to robot behavior. We begin by first defining the semantic task representation used by our grounding models that comes directly from the OO-MDP propositional functions of the domain. Next, we examine our novel DRAGGN framework for language grounding and, in particular, address the separate paths taken by action-oriented and goal-oriented commands through the system as seen in Figure 2. Finally, we discuss two different

| Action-Oriented | Goal-Oriented |
| --- | --- |
| goUp(numSteps) | agentInRoom(room) |
| goDown(numSteps) | blockInRoom(room) |
| goLeft(numSteps) | |
| goRight(numSteps) | |

Table 1: Set of action-oriented and goal-oriented callable units that can be generated by our DRAGGN models in the Cleanup World domain.

implementations of the DRAGGN framework that make different assumptions about the relationship between tasks and constraints. Specifically, we introduce the Joint-DRAGGN (J-DRAGGN), that assumes a probabilistic dependence between tasks (i.e. goUp) and the corresponding arguments (i.e. 5 steps) based on a natural language instruction, and the Independent-DRAGGN (I-DRAGGN) that treats tasks and arguments as independent given a natural language instruction.

### 4.1 Semantic Representation

In order to map arbitrary natural language instructions to either action trajectories or goal conditions, we require a compact but sufficiently expressive semantic representation for both. To this end, we define the *callable unit*, which takes the form of a single-argument function. These functions are paired with *binding arguments* whose possible values depend on the callable unit type. As in MacGlashan et al. (2015) and Arumugam et al. (2017), our approach generates reward function templates, or *lifted* reward functions, for goal-oriented tasks along with environment-specific constraints. Once these templates and constraints are resolved to get a grounded reward function, the associated goal-oriented tasks can be solved by an off-the-shelf planner thereby improving transfer and generalization capabilities.

Goal-oriented callable units (lifted reward functions) are paired with binding arguments that specify properties of environment entities that must be satisfied in order to achieve the goal. These binding arguments are later resolved by the Grounding Module (see Section 4.5) to produce grounded reward functions (OO-MDP propositional logic functions) that are handled by an MDP planner.

Action-oriented callable units directly correspond to the primitive actions available to the robot and are paired with binding arguments defining the number of sequential executions of that action. The full set of callable units along with req-

uisite binding arguments is shown in Table 1.

### 4.2 Deep Recurrent Action/Goal Grounding Network (DRAGGN)

While the Single-RNN model of Arumugam et al. (2017) is effective, it cannot model the compositional argument structure of language. A unit-argument pair not observed at training time will not be predicted from input data, even if the constituent pieces were observed separately. Additionally, the Single-RNN model requires every possible unit-argument pair to be enumerated, to form the output space. As the environment grows to include more objects with richer attributes, this output space becomes intractable.

To resolve this, we introduce the Deep Recurrent Action/Goal Grounding Network (DRAGGN) framework. Unlike previous approaches, the DRAGGN framework maps natural language instructions to *separate* distributions over callable units and (possibly multiple) binding constraints, generating either action sequences or goal conditions. By treating callable units and binding arguments as separate entities, we circumvent the combinatorial dependence on the size of the domain.

This unit-argument separation is inspired by the Neural Programmer-Interpreter (NPI) of Reed and de Freitas (2016). The callable units output by DRAGGN are analogous to the subprograms output by NPI. Additionally, both NPI and DRAGGN allow for subprograms/callable units with an arbitrary number of arguments (by adding a corresponding number of Binding Argument Networks, as shown at the top right of Figure 3a, each with its own output space).

We assume that each natural language instruction can be represented by a single unit-argument pair with only one argument. Consequently, in our experiments, we assume that sentences specifying sequences of commands have been segmented, and each segment is given to the model one at a time. The limitation to a single argument only arises because of the domain's simplicity; as mentioned above, it is straightforward to extend our models to handle extra arguments by adding extra Binding Argument Networks.

To formalize the DRAGGN objective, consider a natural language instruction $l$. Our goal is to find the callable unit $\hat{c}$ and binding arguments $\hat{a}$ that

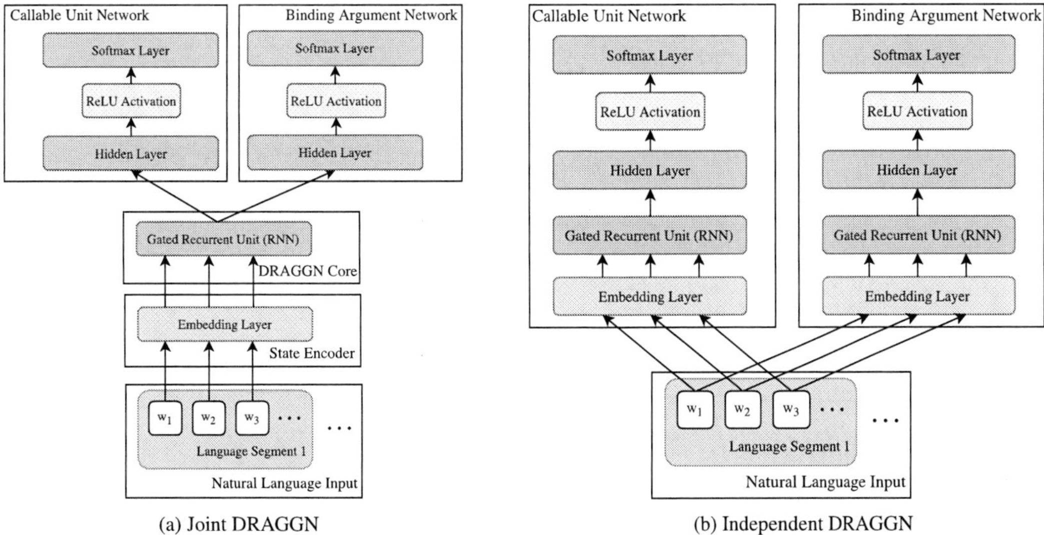



(a) Joint DRAGGN    (b) Independent DRAGGN

Figure 3: Architecture diagrams for the two Deep Recurrent Action/Goal Grounding Network (DRAGGN) models, introduced in Sections 4.3 and 4.4. Both architectures ground arbitrary natural language instructions to callable units (either actions or lifted reward functions), and binding arguments.

maximize the following joint probability:

$$\hat{c}, \hat{\mathbf{a}} = \arg\max_{c,\mathbf{a}} \Pr(c, \mathbf{a} \mid l) \qquad (1)$$

Depending on the assumptions made about the relationship between callable units $c$ and binding arguments $\mathbf{a}$, we can decompose the above objective in two ways: preserving the dependence between the two, and learning the relationship between the units and arguments jointly, and treating the two as independent. These two decompositions result in the Joint-DRAGGN and Independent-DRAGGN models respectively.

Given the training dataset of natural language and the space of unit-argument pairs, we train our DRAGGN models end-to-end by minimizing the sum of the cross-entropy losses between the predicted distributions and true labels for each separate distribution (*i.e.* over callable units and binding arguments). At inference time, we first choose the callable unit with the highest probability given the natural language instruction. We then choose the binding argument(s) with highest probability from the set of valid arguments. The validity of a binding argument given a callable unit is given *a priori*, by the specific environment, rather than being learned at training time.

Our models were trained using Adam (Kingma and Ba, 2014), for 125 epochs, with a batch size of 16, and a learning rate of 0.0001.

## 4.3 Joint DRAGGN (J-DRAGGN)

The Joint DRAGGN (J-DRAGGN) models the joint probability in Equation 1, coupled via the shared RNN state in the DRAGGN Core (as depicted in Figure 3a), but selects the optimizer sequentially, as follows:

$$\hat{c}, \hat{\mathbf{a}} = \arg\max_{c,\mathbf{a}} \Pr(c, \mathbf{a} \mid l) \qquad (2)$$
$$\approx \arg\max_{\mathbf{a}} \left[ \arg\max_{c} \Pr(c, \mathbf{a} \mid l) \right]$$

We first encode the constituent words of our natural language segment into fixed-size embedding vectors. From there, the sequence of word embeddings is fed through an RNN denoted by the DRAGGN Core[1]. After processing the entire segment, the current gated recurrent unit (GRU) hidden state is then treated as a representative vector for the entire natural language segment. This single hidden core vector is then passed to both the Callable Unit Network and the Binding Argument Network, allowing for both networks to be trained jointly, enforcing a dependence between the two.

The Callable Unit Network is a two-layer feed-forward network using rectified linear unit (ReLU) activation. It takes the DRAGGN Core output

---

[1] We use the gated recurrent unit (GRU) as our RNN cell, because of its effectiveness in natural language processing tasks, such as machine translation (Cho et al., 2014), while requiring fewer parameters than the LSTM cell (Hochreiter and Schmidhuber, 1997).

vector as input to produce a softmax probability distribution over all possible callable units. The Binding Argument Network is a separate network with an identical architecture and takes the same input, but instead produces a probability distribution over all possible binding arguments. The two models do not need to share the same architecture; for example, callable units with multiple arguments require multiple different argument networks, one for each possible binding constraint.

### 4.4 Independent DRAGGN (I-DRAGGN)

The Independent DRAGGN (I-DRAGGN), contrary to the Joint DRAGGN, decomposes the objective from Equation 1 by treating callable units and binding arguments as being independent, given the original natural language instruction. More precisely, the I-DRAGGN objective is:

$$\hat{c}, \hat{\mathbf{a}} = \arg\max_{c, \mathbf{a}} \Pr(c \mid l) \Pr(\mathbf{a} \mid l) \qquad (3)$$

The I-DRAGGN network architecture is shown in Figure 3b. Beyond the difference in objective functions, there is another key difference between the I-DRAGGN and J-DRAGGN architectures. Rather than encoding the constituent words of the natural language instruction once, and feeding the resulting embeddings through a DRAGGN Core to generate a shared core vector, the I-DRAGGN model embeds and encodes the natural language instruction *twice*, using two separate embedding matrices and GRUs, one each for the callable unit and binding argument. In this way, the I-DRAGGN model encapsulates two disjoint neural networks, each with their own individual parameter sets that are trained independently. The latter half of each individual network (the Callable Unit Network and Binding Argument Network) remains the same as that of the J-DRAGGN.

### 4.5 Grounding Module

If a goal-oriented callable unit is returned (*i.e.* a lifted reward function), we require an additional step of completing the reward function with environment-specific variables. As described in Arumugam et al. (2017), we use a Grounding Module to perform this step. The Grounding Module maps the inferred callable unit and binding argument(s) to a final grounded reward function that can be passed to an MDP planner. In our implementation, the Grounding Module is a lookup table mapping specific binding arguments to room

| Natural Language | Callable Unit | Argument |
| --- | --- | --- |
| Go to the red room. | agentInRoom | roomIsRed |
| Put the block in the green room. | blockInRoom | roomIsGreen |
| Go up three spaces. | goUp | 3 |

Table 2: Examples of natural language phrases and corresponding callable units and arguments.

ID tokens. A more advanced implementation of the Grounding Module would be required in order to handle domains with non-unique binding arguments (*e.g.* resolving between multiple objects with overlapping attributes).

## 5 Experiments

We assess the effectiveness of both our J-DRAGGN and I-DRAGGN models via instruction grounding accuracy for robot navigation and mobile-manipulation tasks. As a baseline, we compare against the state-of-the-art Single-RNN model introduced by Arumugam et al. (2017).

### 5.1 Procedure

To conduct our evaluation, we use the dataset of natural language commands for the single instance of Cleanup World domain seen in Figure 1, from Arumugam et al. (2017). In the user study, Amazon Mechanical Turk users were presented with trajectory demonstrations of a robot completing various navigation and object manipulation tasks. Users were prompted to provide natural language commands that they believed would have generated the observed behavior. Since the original dataset was compiled for analyzing the hierarchical nature of language, we were easily able to filter the commands down to only those using high-level goal specifications and low-level trajectory specifications. This resulted in a dataset of 3734 natural language commands total.

To produce a dataset of action-specifying callable units, experts annotated low-level trajectory specifications from the Arumugam et al. (2017) dataset. For example, the command "Down three paces, then up two paces, finally left four paces" was segmented into "down three spaces," "then up two paces," "finally left four paces," and was given a corresponding execution trace of goDown 3, goUp 2, goLeft 4. The existing set of grounded reward functions in the dataset were converted to callable units and binding arguments. Examples of both types of language are presented

|  | Action-Oriented | Goal-Oriented | Action-Oriented (Unseen) | Overall |
| --- | --- | --- | --- | --- |
| Single-RNN | $95.8 \pm 0.1\%$ | *$87.2 \pm 0.9\%$* | $0.0 + 0\%$ | $80.0 \pm 0.2\%$ |
| J-DRAGGN | $96.6 \pm 0.2\%$ | *$87.9 \pm 1.9\%$* | $20.2 \pm 20.4\%$ | $83.7 \pm 2.8\%$ |
| I-DRAGGN | $\mathbf{97.0 \pm 0.2\%}$ | $84.9 \pm 1.8\%$ | $\mathbf{97.0 + 0.0\%}$ | $\mathbf{94.7 \pm 0.5\%}$ |

Table 3: Action-oriented and goal-oriented accuracy results (mean and standard deviation across 3 random initializations) on both the standard and unseen datasets. **Bold** indicates the singular model that performed the best on the given task, whereas *italics* denotes the best models that were within the margin of error of each other for the given task. The overall column was computed by taking an average of individual task accuracies, weighted by the number of test examples per task.

in Table 2 with their corresponding callable unit and binding arguments.

To fully show the capabilities of our model, we tested on two separate versions of the dataset. The first is the standard dataset, consisting of a 90-10 split of the collected action-oriented and goal-oriented commands We also evaluated our models on an "unseen" dataset, which consists of a specific train-test split that evaluates how well models can predict previously unseen action sequence combinations. For example, in this dataset the training data might consist only of action sequences of the form goUp 3, and goDown 4, while the test data would only consist of the "unseen" action sequence goUp 4. Note that in both datasets, we assume that the test environment is configured the same as the train environment.

## 5.2 Results

Language grounding accuracies for our two DRAGGN models, as well as the baseline Single-RNN, are presented in Table 3. All three models received the same set of training data, consisting of 2660 low-level action-oriented segments and 693 high-level goal-based sentences. All together, there are 17 unique combinations action-oriented callable units and respective binding arguments, and 6 unique combinations of goal-oriented callable units and binding arguments present in the data. Then, we evaluated all three models on the same set of held-out data, which consisted of 295 low-level segments and 86 high-level sentences.

In aggregate, the models that use callable units for both action- and goal-based language grounding demonstrate superior performance to the Single-RNN baseline, largely due to their ability to generalize, and output combinations unseen at train time. We break down the performance on

each task in the following three sections.

## 5.3 Action Prediction

We evaluate the performance of our models on low-level language that directly specifies an action trajectory. An instruction is correctly grounded if the output trajectory specification corresponds to the ground-truth action sequence. To ensure fairness, we augment the output space of Single-RNN to include all distinct action trajectories found in the training data (an additional 17 classes, as mentioned previously).

All models perform generally well on this task, with Single-RNN correctly identifying the correct action callable unit on 95.8% of test samples, while both DRAGGN models slightly outperform with on 96.6% and 97.0% respectively.

## 5.4 Goal Prediction

In addition to the action-oriented results, we evaluate the ability for each model to ground goal-based commands. An instruction is correctly grounded if the output of the grounding module corresponds to the ground-truth (grounded) reward function.

In our domain, all models predict the correct grounded reward function with an accuracy of 84.9% or higher, with the Single-RNN and J-DRAGGN models being too close to call.

## 5.5 Unseen Action Prediction

The Single-RNN baseline model is completely unable to produce unit-argument pairs that were never seen during training, whereas both DRAGGN models demonstrate some capacity for generalization. The I-DRAGGN model in particular demonstrates a strong understanding of each token within the original natural language utterances which, in large part, comes from the separate embedding spaces maintained for callable units and binding constraints respectively.

# 6 Discussion

Our experiments show that the DRAGGN models have a clear advantage over the existing state-of-the-art in grounding action-oriented language. Furthermore, due to the factored nature of the output, I-DRAGGN generalizes well to unseen combinations of callable units and binding arguments.

Nevertheless, I-DRAGGN did not perform as well as Single-RNN and J-DRAGGN on goal-oriented language. This is possibly due to the small number of goal types in the dataset and the strong overlap in goal-oriented language. Whereas the Single-RNN and J-DRAGGN architectures may experience some positive transfer of information (due to the shared parameters in each of the two models), the I-DRAGGN model does not because of its assumed independence between callable units and binding arguments. This ability to allow for positive information transfer suggests that J-DRAGGN would perform best in environments where there is a strong overlap in the instructional language, with a relatively smaller but complex set of possible action sequences and goal conditions.

On action-oriented language, J-DRAGGN has grounding accuracy of around 20.2% while I-DRAGGN achieves a near-perfect 97.0%. Since J-DRAGGN only encodes the input language instruction once, the resulting vector representation is forced to characterize both callable unit and binding argument features. While this can result in positive information transfer and improve grounding accuracy in some cases (*e.g.* goal-based language), this enforced correlation heavily biases the model towards predicting combinations it has seen before. By learning separate representations for callable units and binding arguments, I-DRAGGN is able to generalize significantly better. This suggests that I-DRAGGN would perform best in situations where the instructional language consists of many disjoint words and phrases.

While our results demonstrate that the DRAGGN framework is effective, more experimentation is needed to fully explore the possibilities and weaknesses of such models. One of the shortcomings in the DRAGGN models is the need for segmented data. We found that all evaluated models were unable to handle long, compositional instructions, such as "Go up three steps, then down two steps, then left five steps". Handling conjunctions of low-level commands requires extending our model to learn how to perform segmentation, or producing sequences of callable units and arguments.

# 7 Conclusion

In this paper, we presented the Deep Recurrent Action/Goal Grounding Network (DRAGGN), a hybrid approach that grounds natural language commands to either action sequences or goal conditions, depending on the language. We presented two separate neural network architectures that can accomplish this task, both of which factor the output space according to the compositional structure of our semantic representation.

We show that overall the DRAGGN models significantly outperform the existing state of the art. Most notably, we show that the DRAGGN models are capable of generalizing to action sequences unseen during training time.

Despite these successes, there are still open challenges with grounding language to novel, unseen environment configurations. Furthermore, we hope to extend our models to handle instructions that are a mixture of goal-oriented and action-oriented language, as well as to long, sequential commands. An instruction such as "go to the blue room, but avoid going through the red hallway" does not map to either an action sequence or a traditional, Markovian reward function. We believe new tools and approaches will need to be developed to handle such instructions, in order to handle the diversity and complexity of human natural language.

# 8 Acknowledgements

This material is based upon work supported by the National Science Foundation under grant number IIS-1637614 and the National Aeronautics and Space Administration under grant number NNX16AR61G.

Lawson L.S. Wong was supported by a Croucher Foundation Fellowship.

# References

Jacob Andreas and Dan Klein. 2015. Alignment-based compositional semantics for instruction following. In *Conference on Empirical Methods in Natural Language Processing*.

Yoav Artzi and Luke Zettlemoyer. 2013. Weakly supervised learning of semantic parsers for mapping instructions to actions. In *Annual Meeting of the Association for Computational Linguistics*.

Dilip Arumugam, Siddharth Karamcheti, Nakul Gopalan, Lawson L.S. Wong, and Stefanie Tellex. 2017. Accurately and efficiently interpreting human-robot instructions of varying granularities. *CoRR* abs/1704.06616.

R. Bellman. 1957. A Markovian decision process. *Indiana University Mathematics Journal* 6:679–684.

Yoshua Bengio, Réjean Ducharme, Pascal Vincent, and Christian Janvin. 2000. A neural probabilistic language model. *Journal of Machine Learning Research* 3:1137–1155.

Kyunghyun Cho, Bart van Merrienboer, Çaglar Gúlçehre, Dzmitry Bahdanau, Fethi Bougares, Holger Schwenk, and Yoshua Bengio. 2014. Learning phrase representations using rnn encoder-decoder for statistical machine translation. In *Empirical Methods in Natural Language Processing*.

Junyoung Chung, Çaglar Gúlçehre, Kyunghyun Cho, and Yoshua Bengio. 2014. Empirical evaluation of gated recurrent neural networks on sequence modeling. *CoRR* abs/1412.3555.

Carlos Diuk, Andre Cohen, and Michael L. Littman. 2008. An object-oriented representation for efficient reinforcement learning. In *International Conference on Machine Learning*.

Juraj Dzifcak, Matthias Scheutz, Chitta Baral, and Paul Schermerhorn. 2009. What to do and how to do it: Translating natural language directives into temporal and dynamic logic representation for goal management and action execution. In *IEEE International Conference on Robotics and Automation*.

Nakul Gopalan, Marie desJardins, Michael L. Littman, James MacGlashan, Shawn Squire, Stefanie Tellex, John Winder, and Lawson L.S. Wong. 2017. Planning with abstract Markov decision processes. In *International Conference on Automated Scheduling and Planning*.

Sachithra Hemachandra, Felix Duvallet, Thomas M. Howard, Nicholas Roy, Anthony Stentz, and Matthew R. Walter. 2015. Learning models for following natural language directions in unknown environments. In *IEEE International Conference on Robotics and Automation*.

Sepp Hochreiter and Jürgen Schmidhuber. 1997. Long short-term memory. *Neural Computation* 9:1735–1780.

Thomas M. Howard, Stefanie Tellex, and Nicholas Roy. 2014. A natural language planner interface for mobile manipulators. In *IEEE International Conference on Robotics and Automation*.

Mohit Iyyer, Varun Manjunatha, Jordan L. Boyd-Graber, and Hal Daumé. 2015. Deep unordered composition rivals syntactic methods for text classification. In *Conference of the Association for Computational Linguistics*.

Diederik P. Kingma and Jimmy Ba. 2014. Adam: A method for stochastic optimization. *CoRR* abs/1412.6980.

James MacGlashan, Monica Babeş-Vroman, Marie desJardins, Michael L. Littman, Smaranda Muresan, Shawn Squire, Stefanie Tellex, Dilip Arumugam, and Lei Yang. 2015. Grounding english commands to reward functions. In *Robotics: Science and Systems*.

Cynthia Matuszek, Evan Herbst, Luke Zettlemoyer, and Dieter Fox. 2012. Learning to parse natural language commands to a robot control system. In *International Symposium on Experimental Robotics*.

Hongyuan Mei, Mohit Bansal, and Matthew R. Walter. 2016. Listen, attend, and walk: Neural mapping of navigational instructions to action sequences. In *AAAI Conference on Artificial Intelligence*.

Tomas Mikolov, Martin Karafiát, Lukás Burget, Jan Cernocký, and Sanjeev Khudanpur. 2010. Recurrent neural network based language model. In *Interspeech*.

Tomas Mikolov, Stefan Kombrink, Lukás Burget, Jan Cernocký, and Sanjeev Khudanpur. 2011. Extensions of recurrent neural network language model. In *IEEE International Conference on Acoustics, Speech, and Signal Processing*.

Rohan Paul, Jacob Arkin, Nicholas Roy, and Thomas M. Howard. 2016. Efficient grounding of abstract spatial concepts for natural language interaction with robot manipulators. In *Robotics: Science and Systems*.

Martin L. Puterman. 1994. Markov decision processes: Discrete stochastic dynamic programming.

Scott E. Reed and Nando de Freitas. 2016. Neural programmer-interpreters. In *International Conference on Learning Representations*.

Stefanie Tellex, Thomas Kollar, Steven Dickerson, Matthew R. Walter, Ashis Gopal Banerjee, Seth Teller, and Nicholas Roy. 2011. Understanding natural language commands for robotic navigation and mobile manipulation. In *AAAI Conference on Artificial Intelligence*.

Adam Vogel and Dan Jurafsky. 2010. Learning to follow navigational directions. In *Annual Meeting of the Association for Computational Linguistics*.

# Are distributional representations ready for the real world?
## Evaluating word vectors for grounded perceptual meaning


**Li Lucy[1]**
lucy3@stanford.edu

**Jon Gauthier[1,2]**
jon@gauthiers.net

[1]Stanford Symbolic Systems    [2]Stanford NLP Group



## Abstract

Distributional word representation methods exploit word co-occurrences to build compact vector encodings of words. While these representations enjoy widespread use in modern natural language processing, it is unclear whether they accurately encode all necessary facets of conceptual meaning. In this paper, we evaluate how well these representations can predict perceptual and conceptual features of concrete concepts, drawing on two semantic norm datasets sourced from human participants. We find that several standard word representations fail to encode many salient perceptual features of concepts, and show that these deficits correlate with word-word similarity prediction errors. Our analyses provide motivation for grounded and embodied language learning approaches, which may help to remedy these deficits.


## 1 Introduction

Distributional approaches to meaning representation have enabled a substantial amount of progress in natural language processing over the past years. They center around a classic insight from at least as early as Harris (1954); Firth (1957):

> You shall know a word by the company it keeps. (Firth, 1957, p. 11)

Popular distributional analysis methods which exploit this intuition such as word2vec (Mikolov et al., 2013) and GloVe (Pennington et al., 2014) have been critical to the success of many recent large-scale natural language processing applications (e.g. Turney and Pantel, 2010; Turian et al., 2010; Collobert and Weston, 2008; Socher et al., 2013; Goldberg, 2016). These methods operationalize distributional meaning via tasks where words are optimized to predict words which co-occur with them in text corpora. These methods yield compact word representations — vectors in some high-dimensional space — which are optimized to solve these prediction tasks. These vector representations form the foundation of practically all modern deep learning models applied within natural language processing.

Despite the success of distributional representations in standard natural language processing tasks, a small but growing consensus within the artificial intelligence community suggests that these methods cannot be sufficient to induce adequate representations of words and concepts (Kiela et al., 2016; Gauthier and Mordatch, 2016; Lazaridou et al., 2015). These sorts of claims, which often draw on experimental evidence from cognitive science (see e.g. Barsalou, 2008), are used to back up arguments for multimodal learning (at the weakest) or complete embodiment (at the strongest). Kiela et al. (2016) claim the following:

> …the best way for acquiring human-level semantics is to have machines learn through (physical) experience: if we want to teach a system the true meaning of "bumping into a wall," we simply have to bump it into walls repeatedly.

Discussions like the one above have an intuitive pull: certainly "bump" is best understood through a sense of touch, just as "loud" is best understood through a sense of sound. It seems inefficient — or perhaps just wrong — to learn these sorts of concepts from distributional evidence.

---

All project code available at github.com/lucy3/grounding-embeddings.




*Proceedings of the First Workshop on Language Grounding for Robotics*, pages 76–85,
Vancouver, Canada, July 30 - August 4, 2017. ©2017 Association for Computational Linguistics

Despite the intuitive pull, there is not much evidence from a computational perspective that grounded or multimodal learning actually earns us anything in terms of general meaning representation. Will our robots and chat-bots be worse off for not having physically bumped into walls before they hold discussions on wall-collisions? Will our representation of the concept *loud* somehow be faulty unless we explicitly associate it with certain decibel levels experienced in the real world? Before we proceed to embed our learning agents in multimodal games and robot-shells, it is important that we have some concrete idea of how grounding actually affects meaning.

This paper presents a thorough analysis of the contents of distributional word representations with respect to this question. Our results suggest that several common distributional word representations may indeed be deficient in the sort of grounded meaning necessary for language-enabled agents deployed in the real world.

## 2 Related work

This paper uses semantic norm datasets to evaluate the content of distributional word representations. Semantic norm datasets consist of concepts and norms concerning their perceptual and conceptual features, as provided by human participants. They are a popular resource within psychology and cognitive science as models of human concept representation, and have been used to explain psycholinguistic phenomena from semantic priming and interference (Vigliocco et al., 2004) to the structure of early word learning in child language acquisition (Hills et al., 2009). Andrews et al. (2009) show how "experiential" semantic norm information can be used to model human judgments of concept similarity. They show that this semantic norm data provides information distinct from the information found in basic word representations. Our work extends the findings of Andrews et al. to a larger semantic norm dataset and evaluates particular implications within natural language processing.

A small NLP literature has compared distributional representations with semantic norm datasets and other external resources. Rubinstein et al. (2015) confirm that word representations are especially effective at predicting taxonomic features versus attributive features. Collell and Moens (2016) find that word representations fail to pre-

| | # word tokens | # word types |
|---|---|---|
| GloVe (Common Crawl) | 840B | 2.2M |
| GloVe (Wiki+Gigaword) | 6B | 400K |
| word2vec | 100B | 3M |

Table 1: Statistics of the corpora used to produce the distributional representations used in this paper.

dict many visual features of concepts, and show how representations from computer vision models can help improve these predictions. Several studies have used distributional representations to reconstruct aspects of these semantic norm datasets (Herbelot and Vecchi, 2015; Fagarasan et al., 2015; Erk, 2016).

The majority of the NLP work in this space has focused on the downstream task of augmenting word representations with novel grounded information, often evaluating on standard semantic similarity datasets (Agirre et al., 2009; Bruni et al., 2012; Faruqui et al., 2015; Bulat et al., 2016). Young et al. (2014) develop an alternative operationalization of denotational meaning using image captioning datasets, and demonstrate gains over distributional representations on textual similarity and entailment datasets.

This applied work has demonstrated that *something* worthwhile is indeed gained by augmenting distributional representations with some orthogonal grounded or multimodal information. We believe it is critical to analyze the original successes and failures of distributional representations in order to motivate this move to grounded meaning representation.

## 3 Meaning representations

### 3.1 Distributional meaning

This paper examines representations produced by two popular unsupervised distributional methods. Table 1 shows the statistics of the corpora used to generate these vectors.

**GloVe:** GloVe (Pennington et al., 2014) estimates word representations $w_i$ by using them to reconstruct a word-word co-occurrence matrix $X$ collected from a large text corpus:

$$L = \sum_{i,j=1}^{V} f(X_{ij}) \left( w_i^T w_j + b_i + b_j - \log X_{ij} \right)^2 \tag{1}$$

| Dataset | # concepts | # features | C/F | F/C |
|---|---|---|---|---|
| McRae | 541 | 2526 | 2.87 | 13.41 |
| CSLB | 638 | 2725 | 3.78 | 16.13 |

Table 2: Semantic norm datasets used in this paper. The final two columns show the mean concepts per feature / features per concept.

here $f(X_{ij})$ is a weighting function on word pairs and $b_i, b_j$ are learned per-word bias terms.

We use two pre-trained GloVe vector datasets: one trained on a concatenation of Wikipedia 2014 and Gigaword 5 (GloVe-WG), and another trained on a Common Crawl dump (GloVe-CC).[1]

**word2vec:** word2vec (Mikolov et al., 2013) estimates word representations by optimizing a skip-gram objective to predict all words $w_j$ within a context window $c$ of a word $w_i$ given their word representations:

$$J = \frac{1}{T} \sum_{i=1}^{T} \sum_{i-c \leq j \leq i+c} \log p(w_j \mid w_i) \quad (2)$$

where $T$ is the total number of words in a corpus. We use a publicly available word2vec dataset trained on the Google News corpus.[2]

### 3.2 Semantic norms

Semantic feature norm datasets consist of reports from human participants about the semantic features of various natural kinds. A proportion of the features contained in these datasets are properties of concepts which may be obvious to humans but are perhaps difficult to find written in text corpora. For this reason, we selected two semantic norm datasets to serve as gold-standard comparisons of concept meaning. Table 2 displays basic statistics about the semantic norm datasets we use in this paper.

**McRae** Our initial experiments use the semantic norm dataset from McRae et al. (2005), which consists of 541 concrete noun concepts with associated feature norms, collected from 725 participants. For a given concept, the McRae dataset includes all feature norms which were reported independently by at least five participants (2,526 in total). After removing concepts indicated to have ambiguous meanings to mitigate

polysemy effects (such as `tank_(army)` and `tank_(container)`) and one concept without a GloVe representation (`dunebuggy`), we had a resulting set of 515 concepts for analysis. The dataset groups features into several perceptual and non-perceptual categories: taxonomic, encyclopedic, function, visual-motion, visual-form_and_-surface, visual-colour, sound, tactile, and taste (McRae et al., 2005). We use the McRae dataset and feature categories to perform basic pilot analyses and form hypotheses about the nature of the distributional representations tested.

**CSLB** We reproduce and extend our results on a second semantic norm dataset collected by the Cambridge Centre for Speech, Language and the Brain (CSLB; Devereux et al., 2014). CSLB contains 638 concepts provided by 123 participants. Their data collection closely followed McRae et al. (2005), though features were included if at least 2 participants named that feature. We removed concepts with two-word names, ambiguous meanings, or missing vector representations to yield a vocabulary of 597 concepts from this dataset. CSLB also includes a feature categorization schema, though the categories are broader than those in McRae: visual perceptual, other perceptual, functional, taxonomic, and encyclopedic.

The mapping between the two categorization schemes is far from perfect. While some perceptual features in McRae are categorized as perceptual features in CSLB, other features (e.g. those related to swimming, flying, eating) are reclassified as "functional" in CSLB. The two datasets disagree on abstract conceptual properties as well. For example, CSLB classifies `is_for_-football` as a functional property, while McRae classifies the comparable feature `associated_-with_football_games` as encyclopedic.

The encyclopedic category is somewhat difficult to distinguish in both datasets. It is composed mainly of abstract factual features, but also contains attributive features such as `is_cold_-blooded` and `does_use_electricity` as well as `is_scary` and `is_cool`.

Meanwhile, the functional category mixes features for behaviors associated with the concept (`does_dive`) as well as functions that people perform on or with the concept (`is_hit`). This classification system may need some readjustments to provide a clear understanding of what is

---

[1] `nlp.stanford.edu/projects/glove`
[2] `code.google.com/archive/p/word2vec`

perceptual and what is conceptual, and it may be that some features, such as `has_a_steering_wheel`, are both.

Given the significant noise of this classification scheme, we focus our investigation on a single contrast between features in clearly perceptual categories (visual, tactile, sound, etc.) and non-perceptual categories (functional and taxonomic). Because the encyclopedic category contains an ambiguous mix of both sorts, we exclude it from our formal predictions later in the paper.

## 4 The feature view

We first investigate how well distributional word representations directly encode information about semantic norms.[3] For each feature in a semantic norm dataset, we construct a binary classification problem which predicts the presence or absence of the feature for each concept. Concretely, for each feature $f_i$ we have a label vector $y_i \in \{0,1\}^{n_c}$, where $n_c$ is the total number of concepts in the dataset, and $y_{ij}$ is 1 when concept $j$ has feature $f_i$ and 0 otherwise. We build label vectors only for features with five or more associated concepts. After filtering, we have $n_f = 267$ label vectors in the McRae dataset and $n_f = 775$ in CSLB.

For each feature, we construct a binary logistic regression model $p^i$ which predicts the presence or absence of the feature for a concept given its word representation $x_j$:

$$p^i(y_{ij} \mid x_j) = \sigma(w_i^T x_j) \qquad (3)$$

This base model is extremely prone to overfitting, as most features have only several associated concepts — that is, each classifier has only a few positive examples — and the input word representations are of a high dimensionality. In order to prevent overfitting, we add an independent L2 regularization term to each regression model. For each feature $f_i$, we use leave-one-out cross-validation to select the regularization parameter $\lambda_i$ which maximizes the following modified logistic objective:

$$
L_i(\lambda_i) = \frac{1}{|f_i|} \sum_{x_j \in f_i} \left( \log p_{-j}^{i,\lambda_i}(y_{ij} = 1 \mid x_j) \right.
$$
$$
\left. + \frac{1}{n_c - |f_i|} \sum_{x_k \notin f_i} \log p_{-j}^{i,\lambda_i}(y_{ik} = 0 \mid x_k) \right)
$$
$$(4)$$

Here $p_{-j}^{i,\lambda_i}(\cdot)$ represents a regression model (Equation (3)) trained without example $(x_j, y_{ij})$ in the training set and with regularization parameter $\lambda_i$. The first term of the summand calculates the log-probability of the left-out concept having the desired feature, and the second term calculates the average log-probability that any other concept (outside of the feature group $f_i$) does not have the feature. The regularization terms $\lambda_i$ are selected independently for each feature to maximize the objective $L_i$.

After fitting the regularized logistic regression models, we calculate a set of "feature fit" metrics. For each feature $f_i$, we evaluate the binary F1 score of its classifier's predictions $p^i(y_i)$. Figure 1 shows each feature as a point in a swarm-plot (grouped by feature category).

Pilot tests with the McRae dataset suggested that the categories associated with strictly perceptual features were not well encoded in the distributional representations relative to strictly non-perceptual categories (taxonomic and functional features).

We use the CSLB dataset as a test set for this prediction. We perform a bootstrap confidence interval test on the difference between the median feature fit scores for CSLB features in non-perceptual and perceptual categories. The 95% confidence intervals on this bootstrap are positive for two of the three representations tested (GloVe-CC and word2vec).[4] Figure 1 shows the feature fit scores on CSLB evaluated with GloVe-CC, and the word2vec evaluation effectively shows the same result: taxonomic and functional features score higher on average than strictly perceptual features. This comparison failed on GloVe-WG, however, where features classed as "functional" scored far lower on average than those in perceptual categories. Across all three sets of distributional representations, the median score of ency-

---

[3]The remainder of this paper describes a general analysis performed on both the McRae and CSLB datasets. We used McRae as a pilot dataset to form hypotheses, and checked these hypotheses on the CSLB dataset as a test set. All of the graphs and numbers reported in this paper correspond to results on CSLB.

[4]GloVe-CC: (7.67%, 24.0%); word2vec: (7.13%, 20.6%); GloVe-WG: (-1.25%, 15.7%).

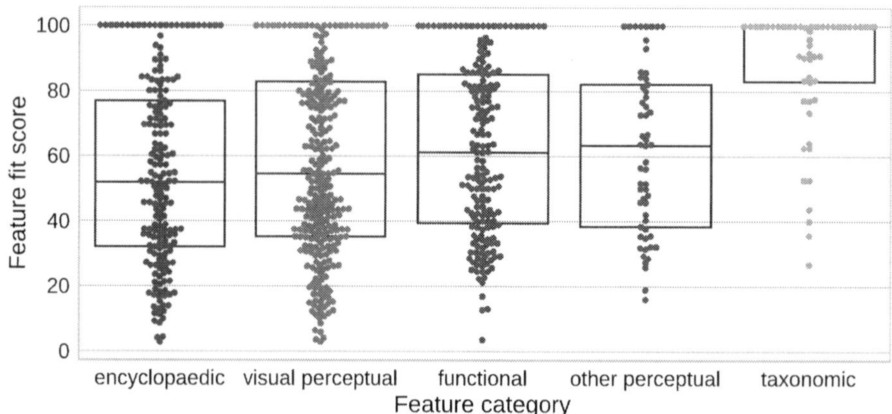


Figure 1: The CSLB feature fit metrics of GloVe-CC, where each point is a feature with at least 5 associated concepts. Feature categories are on the horizontal axis.

| category | feature fit $< 50\%$ | feature fit $> 50\%$ |
|---|---|---|
| other perceptual | `is_chewy, is_solid, is_high_pitched` | `is_hard, does_smell_good_nice, is_juicy` |
| visual perceptual | `is_triangular, has_a_string, is_curved,` | `has_a_clasp, has_a_shell, has_whiskers` |
| encyclopedic | `is_collectable, is_powerful, made_of_-`<br>`tissue` | `is_formal, does_not_fly, is_kept_in_a_-`<br>`cage` |
| functional | `is_roasted, is_for_weddings, is_carried` | `does_shelter, does_chop, is_eaten_edible` |
| taxonomic | `is_a_home, is_a_vessel, is_an_ingredient` | `is_seafood, is_a_boat, is_a_tool` |

Table 3: Examples of features in each category with feature fit scores based on using GloVe-CC to predict norms from CSLB.

clopedic features was well below all other feature categories.

It is obvious from Figure 1 that each category contains a wide range of feature fit values. As discussed earlier in Section 3.2, this categorization of features is far from perfect. Many of the lower-scoring features classed as "encyclopedic" are simple attributive features not deserving of the category label, such as `is_fresh` and `is_filling`. Many of the higher-scoring encyclopedic features seem genuinely encyclopedic, such as `is_found_on_farms`; other high-scoring features are arguably "functional," such as `does_grow_on_trees`. Many of the higher scoring visual perceptual features state structural part-whole relations, such as `has_legs` and `has_an_engine`.

Table 3 provides more examples of low- and high-scoring features in each category. Despite the rather noisy classification scheme used in this dataset, we still managed to find a regular trend in two of three evaluations, matching our expectations from prior pilot experiments. We believe that a revised classification scheme could help to

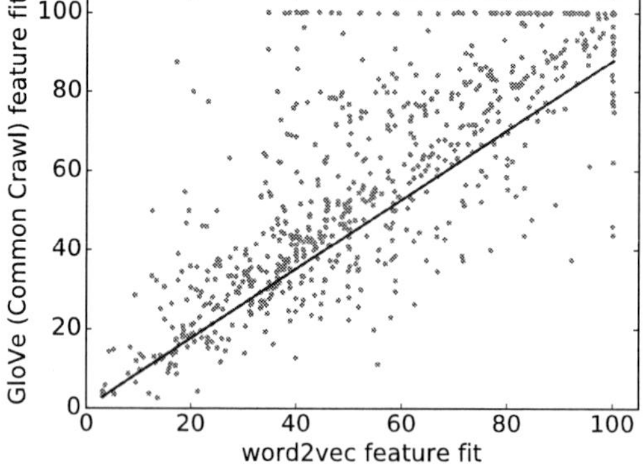


Figure 2: A comparison of CSLB feature fit scores for word2vec and GloVe-CC. Slope: 0.8773; Pearson $r$: 0.8260.

demonstrate a clear difference between perceptual and non-perceptual features in all three datasets.

## 4.1 Matching word representation sources

For each feature, we compare its feature fit score evaluated with GloVe-CC word vectors and its score evaluated with word2vec vectors in Figure 2.

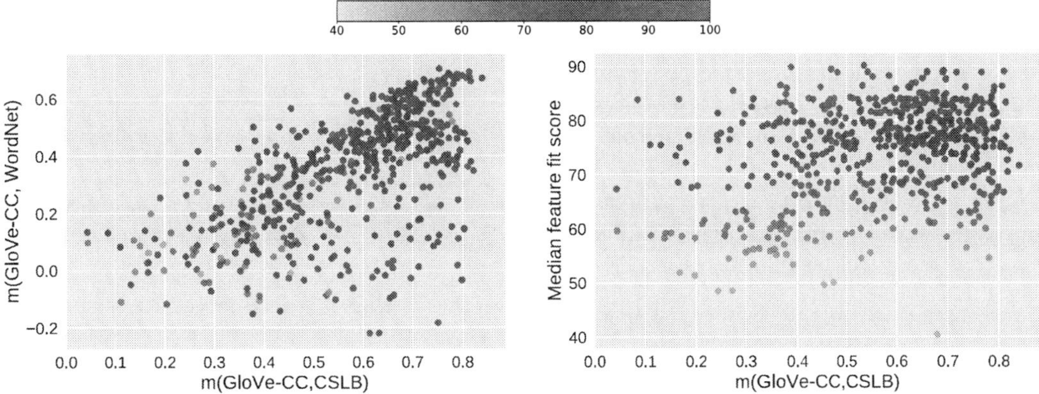

(a) Concepts plotted according to the correlation between their GloVe-CC embeddings and two other sources (CSLB, horizontal axis; WordNet, vertical axis). The points are colored according to their feature fit.

(b) A direct comparison of $m(\text{GloVe-CC}, \text{CSLB})$ (horizontal axis) and the median feature fit scores associated with concept (vertical axis). See main text for statistical tests.

Figure 3: Concept view results.

The trend in the figure suggests that both representations have similar feature fit deficiencies and strengths, though the trend becomes weaker near the $(100\%, 100\%)$ corner — the two representations correlate well at low feature fit scores, and seem to fan out at higher scores. A large group of points also sit in the figure at $y = 100$ and $x = 100$; these features are perfectly captured by one representation and not by the other.

This correlation is somewhat surprising, given that the word2vec and GloVe vectors are the products of different algorithms executed on very different corpora. There are two likely explanations behind this correlation:

1. Some features in the CSLB semantic norm data are unusually difficult, or are perhaps missing associated concepts. GloVe and word2vec correlate in performance because they don't match these noisy or incomplete features.

2. There are systematic deficiencies in the word vectors due to their shared reliance on the distributional method.

It is difficult to differentiate these two explanations on these small semantic norm datasets, but we hope to distinguish these in the future by testing new predictions for concepts not covered in these datasets. We will return to this idea in the conclusion of the paper.

## 5  The concept view

The previous section demonstrated that several classes of perceptual features are not well encoded on average by distributional word representations, and that these deficiencies systematically match across representations. How does this deficiency in feature representation carry over into computations on the word representations themselves?

We evaluate the matching between distributional representations and representations from other sources by comparing their predictions of word-word similarity. For distributional word representations, we compute word-word similarity by cosine distance:

$$\text{sim}(i, j) = \cos(x_i, x_j) \qquad (5)$$

We derive compact concept representations from the semantic norm datasets with LSA (Landauer et al., 1998). We compute a truncated SVD on the feature matrix $Y \in \{0, 1\}^{n_c \times n_f}$, which is the concatenation of the binary feature label vectors introduced in Section 4. We define concept-concept similarity by the cosine distance between their corresponding LSA vectors.

As a secondary data source, we also compute word-word similarity judgments from the Word-Net taxonomy (Miller, 1995). We use the Resnik metric (Resnik et al., 1999) to compute the similarity between concept names $c_i, c_j$:

$$\text{sim}_{\text{resnik}}(c_i, c_j) = \max_{c \in S(c_i, c_j)} - \log p(c) \qquad (6)$$

where $S(c_i, c_j)$ selects the common ancestors of the concepts in the WordNet taxonomy, and $p(c)$ is the unigram probability of a concept as computed on an external corpus. This selects the ancestor of the two concepts in the taxonomy which has maximal information content (surprisal). We use WordNet as additional verification that the trends observed between semantic norms and distributional representations are non-coincidental.

We use these similarity metrics to compute pairwise distance measures for concepts present in the semantic norm datasets. For each metric, we produce a symmetric pairwise distance matrix $D \in \mathbb{R}^{n_c \times n_c}$, where an element $D_{ij}$ indicates the distance between concepts $i$ and $j$ according to the metric.

We next compute how well each concept's pairwise similarity is correlated between the various metrics. For a given concept, we compute the Pearson correlation between the concept's GloVe/word2vec pairwise distance vector and the LSA and WordNet pairwise distance vectors.[5] The correlation values of interest are $m(\text{GloVe/word2vec}, \text{CSLB})$ and $m(\text{GloVe/word2vec}, \text{WordNet})$ — that is, the correlations between the pairwise distance vectors for GloVe/word2vec and CSLB and between the pairwise distance vectors for GloVe/word2vec and WordNet.

Figure 3a plots both of these correlation values evaluated with GloVe-CC for all concepts. The two $m$ measures are evidently positively correlated, though with some noise ($r = 0.6160$). This is to be expected, as the CSLB dataset and WordNet overlap only partially in the semantic features they encode.

Each concept in Figure 3 is colored according to the median feature fit score of its associated features. In Figure 3b, we show this feature fit metric on the vertical axis. There is a positive relationship here between feature fit scores and the correlation metric $m(\text{GloVe-CC}, \text{CSLB})$ ($r = 0.3323$). Because the correlation between $m(\cdot, \text{CSLB})$ and feature fit metrics is weaker than expected, we run post-hoc multiple regression significance tests for each distributional representation. An F-test shows that the regression feature $m(\cdot, \text{CSLB})$ significantly improves predictions of feature fit val-

---

[5]The Pearson correlation between two vectors is equivalent to the cosine distance between their mean-centered forms.

| Domain | Feature fit | Concepts |
|---|---|---|
| 10 | 61.03% | bread, cheese, chocolate, coffee, glue, ham, jam, jelly, ketchup, moss, soup, tea, yoghurt |
| 15 | 67.18% | artichoke, asparagus, aubergine, bean, cabbage, flour, gherkin, leek, mango, pineapple, potato, pumpkin, rhubarb, seaweed |
| 25 | 75.62% | bouquet, buttercup, carnation, daffodil, daisy, dandelion, fern, geranium, hyacinth, lily, marigold, orchid, pansy, poppy, rose, sunflower, tulip |
| 31 | 78.22% | bayonet, bomb, cannon, crossbow, dagger, grenade, gun, pistol, revolver, rifle, shotgun, sword |
| 36 | 82.18% | book, catalogue, menu, dictionary, encyclopaedia, textbook |

Table 4: Selected domains from the clustering analysis on GloVe-CC, with median feature fit scores over concepts.

ues relative to a baseline model for all three representations[6,7].

There is substantial variance in the predictions of the distributional representations due to factors outside of the scope of the semantic norm data. The mismatch in predictions between distributional representations is nevertheless a statistically significant predictor of feature fit metrics. This suggests that the feature-level deficiencies discovered in the previous section have concrete implications in terms of word-word similarity measures.

### 5.1 Domain-level analysis

We next investigate whether some domains of concepts are particularly affected by the deficiencies discussed in the previous sections. We perform agglomerative clustering on concepts from the CSLB dataset using a custom distance metric:

$$d(i, j) = ||\text{LSA}_i - \text{LSA}_j||_2 + \alpha(\text{FF}_i - \text{FF}_j)^2 \quad (7)$$

where $\text{LSA}_i$ is the LSA vector representation computed from the semantic norm data for concept $i$ as introduced earlier in this section, and $\text{FF}_i$ is the median feature-fit score for a concept $i$. We select the weight $\alpha$ manually to produce the most semantically coherent clusters.

---

[6]The baseline regression model predicts a concept's feature fit from these baseline features: log(word frequency in Brown corpus), log(# associated features), log(total # feature reports for the concept), # WordNet senses.

[7]GloVe-CC: $F^* = 41.297, p < 10^{-9}$; GloVe-WG: $F^* = 68.783, p < 10^{-15}$, word2vec: $F^* = 41.27, p < 10^{-9}$

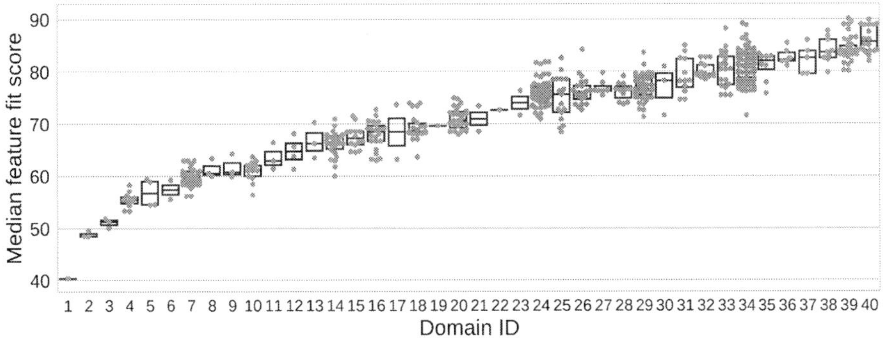

Figure 4: Concept domains derived from the CSLB semantic norm data. Each point represents a concept. The vertical axis is the median feature fit score of the concept's features on GloVe-CC.

Figure 4 shows the distribution of feature fit scores for each of the resulting 40 domains. We find that settings of $\alpha$ which yield semantically coherent clusters also yield groups of concepts with very low variance in feature fit scores. In Table 4 we list select domains and their median feature fit scores. This clustering suggests that deficiencies at the feature level affect entire coherent semantic domains of concepts.

## 6  Conclusion

This paper has analyzed how well various standard distributional representations encode aspects of grounded meaning. We chose to use semantic norm datasets as a gold standard of grounded meaning, and tested how word representations predicted features within these datasets. We grouped these features into high-level categories and found that, despite large within-category variance, several standard distributional representations underperformed on average in predicting perceptual features. The difference in prediction performance proved statistically significant on two of the three representations we evaluated. These deficiencies in feature encoding matched between GloVe and word2vec representations trained on different corpora, suggesting that certain classes of features may be poorly represented by distributional methods in general.

We also examined the consequences of these deficiencies in feature encoding for the word representations themselves. We compared the word-word similarity predictions made with distributional representations with those made with the semantic norm dataset and with WordNet, and found that words having features badly encoded within the distributional representations were also likely to make different similarity predictions than the predictions from these two corpora. A final domain-level concept analysis suggested that some semantic domains are particularly impacted by these issues in feature encoding.

The semantic norm datasets used in this paper are subject to saliency biases: they only contain the concept-feature mappings which experimental subjects think to mention when queried. These saliency effects add noise to our results, as mentioned in Section 4.1, and may have caused us to generally underestimate the performance of distributional models within all feature categories. In future work, we plan to repeat the sorts of tests conducted in this paper while avoiding possible saliency confounds. We also plan to develop a causal explanation for the deficiencies in the word embeddings found in this paper, showing how co-occurrence information (or lack thereof) present in the training corpus can bias performance on these tasks. Both of these studies will verify that the results we have found are due entirely to deficiencies in distributional methods rather than in the datasets used here.

We think these deficiencies should be worrying: if neural models of language are to have any knowledge about concepts, it ought to be in their word embeddings. Our findings show that these embeddings are lacking in basic features of perceptual meaning. These results suggest that distributional meaning (as operationalized by modern distributional models) may miss out on fundamental elements of semantics. We hope they will help motivate further work in developing multimodal representations which can prepare us to deploy more fluent language agents in the real world.

**Acknowledgements**

We thank Christopher D. Manning, Peng Qi, Pakapol Supaniratisai, Keenon Werling, and members of the Stanford, University of Washington, and Berkeley NLP communities for useful discussions, and the anonymous reviewers for their insightful comments.

**References**

Eneko Agirre, Enrique Alfonseca, Keith B. Hall, Jana Kravalova, Marius Pasca, and Aitor Soroa. 2009. A Study on Similarity and Relatedness Using Distributional and WordNet-based Approaches. In *HLT-NAACL*.

Mark Andrews, Gabriella Vigliocco, and David Vinson. 2009. Integrating experiential and distributional data to learn semantic representations. *Psychological review* 116(3):463.

Lawrence W Barsalou. 2008. Grounded cognition. *Annu. Rev. Psychol.* 59:617–645.

Elia Bruni, Gemma Boleda, Marco Baroni, and Nam-Khanh Tran. 2012. Distributional Semantics in Technicolor. In *Proceedings of the 50th Annual Meeting of the Association for Computational Linguistics: Long Papers - Volume 1*. Association for Computational Linguistics, Stroudsburg, PA, USA, ACL '12, pages 136–145. http://dl.acm.org/citation.cfm?id=2390524.2390544.

Luana Bulat, Douwe Kiela, and Stephen Clark. 2016. Vision and feature norms: Improving automatic feature norm learning through cross-modal maps. In *HLT-NAACL*.

Guillem Collell and Marie-Francine Moens. 2016. Is an image worth more than a thousand words? on the fine-grain semantic differences between visual and linguistic representations. In *Proceedings of COLING 2016, the 26th International Conference on Computational Linguistics: Technical Papers*. The COLING 2016 Organizing Committee, Osaka, Japan, pages 2807–2817. http://aclweb.org/anthology/C16-1264.

Ronan Collobert and Jason Weston. 2008. A unified architecture for natural language processing: Deep neural networks with multitask learning. In *Proceedings of the 25th international conference on Machine learning*. ACM, pages 160–167.

Barry J Devereux, Lorraine K Tyler, Jeroen Geertzen, and Billi Randall. 2014. The centre for speech, language and the brain (cslb) concept property norms. *Behavior research methods* 46(4):1119–1127.

Katrin Erk. 2016. What do you know about an alligator when you know the company it keeps? *Semantics and Pragmatics* 9(17):1–63. https://doi.org/10.3765/sp.9.17.

Luana Fagarasan, Eva Maria Vecchi, and Stephen Clark. 2015. From distributional semantics to feature norms: grounding semantic models in human perceptual data. In *IWCS*.

Manaal Faruqui, Jesse Dodge, Sujay Kumar Jauhar, Chris Dyer, and Eduard Hovy. 2015. Retrofitting Word Vectors to Semantic Lexicons .

John Rupert Firth. 1957. A synopsis of linguistic theory, 1930-1955. *Studies in linguistic analysis* .

Jon Gauthier and Igor Mordatch. 2016. A paradigm for situated and goal-driven language learning. *arXiv preprint arXiv:1610.03585* .

Yoav Goldberg. 2016. A primer on neural network models for natural language processing. *Journal of Artificial Intelligence Research* 57:345–420.

Zellig S Harris. 1954. Distributional structure. *Word* 10(2-3):146–162.

Aurélie Herbelot and Eva Maria Vecchi. 2015. Building a shared world: mapping distributional to model-theoretic semantic spaces. In *EMNLP*.

Thomas T Hills, Mounir Maouene, Josita Maouene, Adam Sheya, and Linda Smith. 2009. Categorical structure among shared features in networks of early-learned nouns. *Cognition* 112(3):381–396.

Douwe Kiela, Luana Bulat, Anita L Vero, and Stephen Clark. 2016. Virtual embodiment: A scalable long-term strategy for artificial intelligence research. *arXiv preprint arXiv:1610.07432* .

Thomas K Landauer, Peter W Foltz, and Darrell Laham. 1998. An introduction to latent semantic analysis. *Discourse processes* 25(2-3):259–284.

Angeliki Lazaridou, Nghia The Pham, and Marco Baroni. 2015. Combining language and vision with a multimodal skip-gram model. In *HLT-NAACL*.

Ken McRae, George S Cree, Mark S Seidenberg, and Chris McNorgan. 2005. Semantic feature production norms for a large set of living and nonliving things. *Behavior research methods* 37(4):547–559.

Tomas Mikolov, Ilya Sutskever, Kai Chen, Greg S Corrado, and Jeff Dean. 2013. Distributed representations of words and phrases and their compositionality. In *Advances in neural information processing systems*. pages 3111–3119.

George A Miller. 1995. Wordnet: a lexical database for english. *Communications of the ACM* 38(11):39–41.

Jeffrey Pennington, Richard Socher, and Christopher D. Manning. 2014. Glove: Global vectors for word representation. In *Empirical Methods in Natural Language Processing (EMNLP)*. pages 1532–1543. http://www.aclweb.org/anthology/D14-1162.

Philip Resnik et al. 1999. Semantic similarity in a taxonomy: An information-based measure and its application to problems of ambiguity in natural language. *J. Artif. Intell. Res.(JAIR)* 11:95–130.

Dana Rubinstein, Effi Levi, Roy Schwartz, and Ari Rappoport. 2015. How Well Do Distributional Models Capture Different Types of Semantic Knowledge? In *ACL*.

Richard Socher, Alex Perelygin, Jean Y Wu, Jason Chuang, Christopher D Manning, Andrew Y Ng, Christopher Potts, et al. 2013. Recursive deep models for semantic compositionality over a sentiment treebank. In *Proceedings of the conference on empirical methods in natural language processing (EMNLP)*. Citeseer, volume 1631, page 1642.

Joseph Turian, Lev Ratinov, and Yoshua Bengio. 2010. Word representations: a simple and general method for semi-supervised learning. In *Proceedings of the 48th annual meeting of the association for computational linguistics*. Association for Computational Linguistics, pages 384–394.

Peter D Turney and Patrick Pantel. 2010. From frequency to meaning: Vector space models of semantics. *Journal of artificial intelligence research* 37:141–188.

Gabriella Vigliocco, David P Vinson, William Lewis, and Merrill F Garrett. 2004. Representing the meanings of object and action words: The featural and unitary semantic space hypothesis. *Cognitive psychology* 48(4):422–488.

Peter Young, Alice Lai, Micah Hodosh, and Julia Hockenmaier. 2014. From image descriptions to visual denotations: New similarity metrics for semantic inference over event descriptions. *Transactions of the Association for Computational Linguistics* 2:67–78.

# Sympathy Begins with a Smile, Intelligence Begins with a Word: Use of Multimodal Features in Spoken Human-Robot Interaction


**Jekaterina Novikova, Christian Dondrup, Ioannis Papaioannou** and **Oliver Lemon**

Interaction Lab
Heriot-Watt University
Edinburgh, EH14 4AS, UK
`{j.novikova, c.dondrup, i.papaioannou, o.lemon}@hw.ac.uk`



## Abstract

Recognition of social signals, from human facial expressions or prosody of speech, is a popular research topic in human-robot interaction studies. There is also a long line of research in the spoken dialogue community that investigates user satisfaction in relation to dialogue characteristics. However, very little research relates a combination of multimodal social signals and language features detected during spoken face-to-face human-robot interaction to the resulting user perception of a robot. In this paper we show how different emotional facial expressions of human users, in combination with prosodic characteristics of human speech and features of human-robot dialogue, correlate with users' impressions of the robot after a conversation. We find that *happiness* in the user's recognised facial expression strongly correlates with likeability of a robot, while *dialogue-related* features (such as number of human turns or number of sentences per robot utterance) correlate with perceiving a robot as intelligent. In addition, we show that facial expression, emotional features, and prosody are better predictors of human ratings related to perceived robot likeability and anthropomorphism, while linguistic and non-linguistic features more often predict perceived robot intelligence and interpretability. As such, these characteristics may in future be used as an online reward signal for in-situ Reinforcement Learning-based adaptive human-robot dialogue systems.

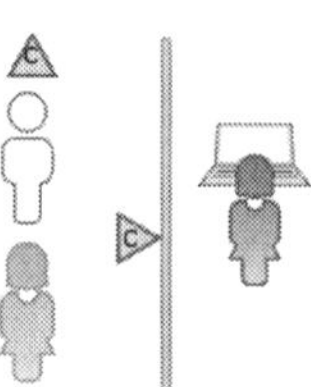



Figure 1: Left: a live view of experimental setup showing a participant interacting with Pepper. Right: a diagram of experimental setup showing the participant (green) and the robot (white) positioned face to face. The scene was recorded by cameras (triangles C) from the robot's perspective focusing on the face of the participant and from the side, showing the whole scene. The experimenter (red) was seated behind a divider.

## 1 Introduction

Social signals, such as emotional expressions, play an important role in human-human interaction, thus they are increasingly recognised as an important factor to be considered both in human-robot interaction research (Cid et al., 2013; Novikova et al., 2015; Devillers et al., 2015) and in the area of spoken dialogue systems (Herm et al., 2008; Meena et al., 2015).

Recognition of human social signals has become a popular topic in Human-Robot Interaction (HRI) in recent years. Social signals are recognized well from human facial expressions or prosodic features of speech (Ekman, 2004; Zeng et al., 2009), and have become the most popular methods for recognising human affective signals in human-robot interaction (Rázuri et al., 2015; Devillers et al., 2015; Cid et al., 2013).

In human-robot interaction, recognized human emotions are mostly used for mimicking human behaviour and enhancing the empathy towards a robot both in children (Tielman et al., 2014) and




*Proceedings of the First Workshop on Language Grounding for Robotics*, pages 86–94,
Vancouver, Canada, July 30 - August 4, 2017. ©2017 Association for Computational Linguistics

in adult users (Tapus and Mataric, 2007).

In the area of spoken dialogue systems, signals recognised from linguistic cues and prosody have been used to detect problematic dialogues (Herm et al., 2008) and to assess dialogue quality as a whole (Schmitt and Ultes, 2015). This type of dialogue-related signals has also been used to automatically detect miscommunication (Meena et al., 2015), or to predict the user satisfaction (Schmitt et al., 2011).

However, there is very little research combining the areas of detecting multi-modal signals during spoken HRI and evaluation of human-robot conversation, and using them to create an adaptive social dialogue.

In this paper, we make a first step towards building a multi-modally-rich, conversational, and human-like robotic agent, potentially able to react to the changes in human behaviour during face-to-face dialogue and able to adjust the dialogue strategy in order to improve an interlocutor's impression. We present a setup that targets the development of a dialogue system to explore verbal and non-verbal conversational cues in a face-to-face situated dialogue with a social robot. We show that different emotional facial expressions of a human interlocutor, in combination with prosodic characteristics of human speech and features of human-robot dialogue, correlate strongly with users' perceptions of a robot after a conversation. Based on these features, we developed a model capable of predicting potential human ratings of a robot and discuss its implications for future work in developing adaptive human-robot dialogue systems.

## 2 Experiment Setup and Evaluation

The human-robot dialogue system was evaluated via a user study in which human subjects interacted with a Pepper robot[1] acting autonomously using the system described in (Papaioannou and Lemon, 2017; Papaioannou et al., 2017). The dialogue system used, combines task-based with chat-based dialogue features, deciding the most appropriate action on each consequent turns, using a pre-trained Reinforcement Learning (RL) policy. The robot decides among a pool of possible actions $a_t \in A$ where $A = $ [*PerformTask, Greet, Goodbye, Chat, GiveDirections, Wait, RequestTask, RequestShop*]. If a task is recognised

in the user utterance (e.g. "where can I find discounts"), a response is synthesized using database lookup and predefined utterances (like the example shown in Table 3). If no task was recognised, then the user request is being forwarded to a Chatbot, written in AIML and based on the chatbot *Rosie*[2], where a chat-style response is formulated based on AIML template/ pattern matching.

All interactions were in English. The physical setup of the experiment can be seen in Figure 1.

### 2.1 Experimental Scenario

The task and the setup chosen in the study were considered as first steps towards understanding how a humanoid social robot should behave in the context of a shopping mall while also providing useful information to the mall's visitors. To this end, participants were asked to imagine that they were entering a shopping mall they had never been to before where the robot was installed in the entry area interacting with visitors one at a time. Participants were asked to complete as many as possible of the following five tasks:

- Get information from the robot on where to get a coffee.

- Get information from the robot on where to buy clothes.

- Get the directions to the clothing shop of their choice.

- Find out if there are any current sales or discounts in the shopping mall and try to get a voucher from the robot.

- Make a selfie with the robot.

Instructions were given to use natural language spontaneously while interacting with the robot.

### 2.2 Participants and Experimental Design

41 people (13 females, 28 males) participated in our study, ranging in age from 18 to 38 (M=24.46, SD=4.72). The majority of them were students (93% students and 7% staff) that had no or little previous experience with robots (56% with little or no experience, 39% with some experience, and 5% with a lot of experience).

Participants were initially given a briefing script describing the goal of the task and providing hints

---

[1] http://doc.aldebaran.com/2-5/home_pepper.html

[2] http://github.com/pandorabots/rosie

on how to better communicate with the robot, e.g. "wait for your turn to speak" and "please keep in mind that the robot only listens to you while its eyes are blinking blue"[3]. We reassured our participants that we were testing the robot, not them, and controlled environment-introduced biases by avoiding non-task-related distractions during the experiment. During experimental sessions, participants stood in front of the robot and the experimenter was hidden in another corner of the room but available in case the participant would need any help (see Figure 1).

At the end of the experiment participants were debriefed and received a £10 gift voucher. The duration of each session did not exceed thirty minutes.

## 2.3  Measured Variables

We collected a range of objective measures from the log files, video and audio recordings of the interactions, and transcripts of dialogues. From the audio recordings, we collected a set of different prosodic and dialogue-related features. From the video recordings, we collected the data on emotional intensities detected based on human facial expressions. From the dialogue transcripts, we collected a set of linguistic features, such as lexical diversity, length of utterance etc.

In addition, we considered a range of subjective measures for a qualitative evaluation. For that, after each interaction session participants were asked to fill in a questionnaire to assess their perception of the robot.

**Emotions** were detected and recognised using the Microsoft Emotion API for Video[4]. This API takes video frames as an input (see Figure 2), and returns the confidence across a set of emotions for the group of faces in the image over a period of time. The emotions detected are *happiness, sadness, surprise, anger, fear, contempt, disgust, or neutral*. Happiness, surprise, and sadness were selected for analysis in this work, because they had the highest average or maximum values across all recorded videos.

**Prosodic Features** used in this work are the following: average fundamental frequency of speech F0, maximum F0, and difference between maximum and minimum F0 values.

Figure 2: Screenshots of the recorded video, showing different facial expressions detected during a dialogue with the robot.

**Non-linguistic Dialogue Features** used in this work contain speech duration (in sec), number of turns, number of completed tasks, number of self-repetitions and a ratio of tasks per turn.

**Linguistic Dialogue Features** used in this work consist of utterance length (in characters), a ratio of words per utterance and unique words per utterance, number of sentences within an utterance, lexical diversity, a ratio of words per sentence and a ratio of unique words per sentence.

**Perception of Robot** was assessed using responses on the questionnaire filled by participants at the end of each interaction session. The questionnaire was based on a combination of the User Experience Questionnaire UEQ (Laugwitz et al., 2008) and the Godspeed Questionnaire (Bartneck et al., 2009). It consisted of 21 pairs of contrasting characteristics that may apply to the robot, and are grouped into four groups of *Anthropomorphism, Likeability, Perceived Intelligence, and User Expectations*. The Anthropomorphism group consists of the following pairs of characteristics: *fake – natural, machinelike – humanlike, unconscious – conscious, artificial – lifelike*. Likeability consists of: *unfriendly – friendly, unkind – kind, unpleasant – pleasant, awful – nice, annoying – enjoyable, disliked – liked*. The group of Perceived Intelligence consists of: *incompetent – competent, ignorant – knowledgeable, irresponsive – responsive, unintelligent – intelligent, foolish – sensible*. The Interpretability group consists of: *does not meet expectations – meets expectations, obstructive – supportive, unpredictable – predictable, confusing – clear, complicated – easy, not understandable – understandable*. Users were asked to evaluate perception of a robot on a 5-point Likert scale, where the minimum value was 1 and the maximum was 5.

The validity of the used questionnaire was tested by measuring its internal consistency with Cronbach's $\alpha$, which was equal to 0.93 (high con-

---

[3]Pepper's default way of communicating that it is listening.

[4]https://www.microsoft.com/cognitive-services/en-us/emotion-api

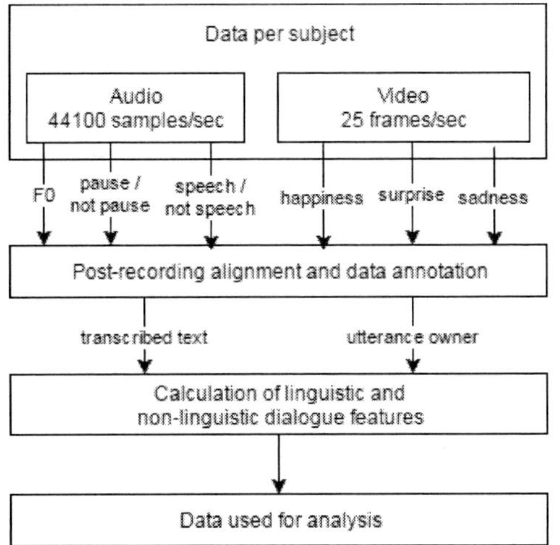


Figure 3: A Flow chart showing the process of synchronising different streams of data, and collecting corresponding parts of data for analysis.

sistency). Based on the high value of the Cronbach's $\alpha$, we assume that that our participants in the given context interpreted the robot characteristics, provided in the questionnaire, in an expected way.

## 3 Multimodal Data Collection and Analysis

Data collected during the experiment required additional processing, alignment and annotation, as shown in Figure 3. Prosodic features of F0, and dialogue-related features showing presence and absence of pauses and presence/absence of speech were collected from audio recordings with a rate of 44100 samples per second. Values of emotional intensities were collected from video recordings with a rate of 25 frames per seconds. All the data was aligned after recording, using average values of prosodic features per frame. Afterwards, data was annotated in ELAN[5] detecting associations between an utterance and its owners. Finally, the dialogue texts were transcribed and linguistic features were calculated using R packages *stringr, stringi, tidytext*, and *qdap*.

A summary of collected data is provided in Table 1. Specifically, the summary results show that the F0 value of human speech changes a lot during the conversation, with a maximum value being more than twice as large as an average value. Average emotional intensities of surprise and sadness, on the other hand, do not differ much

[5]https://tla.mpi.nl/tools/tla-tools/elan

| Group of features | Feature | Human | Robot |
|---|---|---|---|
| Emotional features | Happiness | 0.40 | NA |
| | Surprise | 0.01 | NA |
| | Sadness | 0.01 | NA |
| Prosodic features | F0, avg | 173.87 | NA |
| | F0, max | 398.63 | NA |
| | F0, diff | 338.49 | NA |
| Linguistic dialogue-related features | Utterance length | 21.34** | **26.78** |
| | Words / Utterance | 5.74** | **7.16** |
| | Unique words / Utterance | 5.49* | **6.70*** |
| | Lexical Diversity | **0.97** | 0.95 |
| | No of sentences | 1.13 | **1.24** |
| | Words / Sentence | 5.19 | **5.99*** |
| | Unique words / Sentence | 5.03 | **5.63** |
| Non-linguistic dialogue-related features | Speech duration, sec | 36.54 | **48.92** |
| | No of turns | **32.81** | 29.38 |
| | No of completed tasks | 4.00 | NA |
| | No of self-repetitions | 3.13 | **4.19** |
| | Tasks / Turn | **0.16** | 0.15 |

Table 1: Descriptive statistics of emotional, prosodic, non-linguistic and linguistic dialogue features for human and robot actors. Here, bold indicates a higher value, ** denotes $p < 0.01$, * denotes $p < 0.05$

($\pm 0.001$), and the maximum values of all the emotional intensities are usually close to 1.

Results of non-linguistic dialogue-related features show that the robot on average speaks significantly longer than humans during a dialogue. Humans tend to have a higher number of turns, although they less frequently repeat themselves. These differences, however, are not significant.

Results of linguistic features reveal more significant differences between robot and human language. For example, the results show that humans on average speak in significantly shorter utterances compared to a robot, both in terms of a number of characters and a number of words per utterance. The robot uses more sentences per utterance on average, although this difference is not significant. The lexical diversity, which was calculated as a ratio of unique words and a total number of words in an utterance, shows a slightly higher value in human language rather than robot's.

Values of linguistic features differ significantly between human and robot language, which leads us to investigate in more details the textual dialogue data in terms of lexical variety and syntactic complexity.

## 4 Linguistic Analysis of Dialogues

Following Gardent et al. (2017), we analyse the dialogue textual data in terms of length of ut-

| Speaker | LS | MSTTR | D-level complexity |
|---|---|---|---|
| robot | 0.44 | **0.61** | **1.71*** |
| human | **0.47*** | 0.59 | 1.68 |

Table 2: Results of linguistic dialogue analysis. * denotes p < 0.05.

terances, lexical richness, and syntactic variation. The results are summarised in Tables 1 and 2 and grouped by a speaker, i.e. robot and human.

## 4.1 Length of utterances

Results presented in Table 1 show that robot utterances are significantly longer than those of their human interlocutors, both in terms of words per utterance and sentences per utterance. This may be partly explained by the fact that a turn-taking process was not very natural and thus was not always successful during the dialogue. It usually took some time for people to learn how to communicate with Pepper properly and to start speaking to the robot only when it was listening. As a result, from time to time people were interrupted by the robot, while they never tried to interrupt the robot themselves. Shorter average length of human utterances is also caused by the way people tend to deal with disfluencies of a dialogue, e.g. rephrasing and shortening their previous utterance in order to emphasise the most important keywords (see an example in Table 3). The robot utterances, on the other hand, were not shortened or changed in any other way in the case of dialogue disfluencies.

HUMAN (H): By the way, I'm a student so I don't have a lot of money. So, is it possible to have some shop with sales or discounts? [30 words]

ROBOT (R): Who, specifically, does? [dialogue disfluency]

H: To have some discounts somewhere. [5 words]

R: There are 2 shops that have sales nearby. These are Tesco, and Phone Heaven.

H: Thank you very much.

Table 3: An example of shortening as a result of dialogue disfluency.

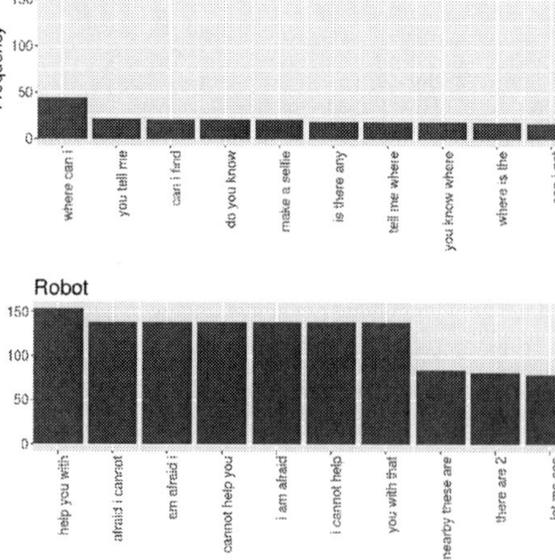


Figure 4: Distribution of the top-10 most frequent trigrams in human and robot language.

## 4.2 Lexical Richness

We used the Lexical Complexity Analyser (Lu, 2009) to measure various dimensions of lexical richness, such as lexical sophistication, lexical diversity and mean segmental type-token ratio. We complement the traditional measure of lexical diversity type-token ratio (TTR) with the more robust measure of mean segmental type-token ratio (MSTTR) (Lu, 2012), which divides all the dialogues into successive segments of a given length and then calculates the average TTR of all segments. The higher the value of MSTTR, the more diverse is the measured text. We also measure *lexical sophistication* (LS), also known as lexical rareness, which is calculated as the proportion of lexical word types not on the list of 2,000 most frequent words generated from the British National Corpus. In addition, we measure *lexical diversity* (LD) as a ratio of unique and total words per utterance.

The results presented in Table 2 show that human utterances, although being significantly shorter, are significantly richer than those of the robot, both in terms of lexical diversity and lexical sophistication. MSTTR values do not differ significantly between human and robot utterances. This leads us to investigate the distribution of frequencies of bigrams and trigrams in human and robot utterances during dialogues.

The majority of both robot (61%) and human (62%) bigrams are only used once in all the dialogues. However, the mean frequency of bigrams

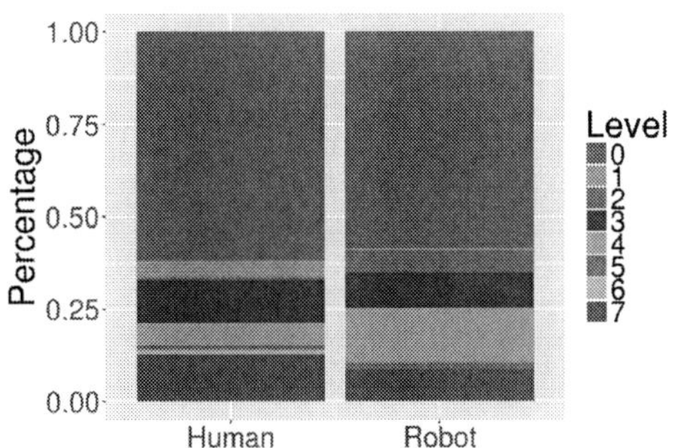

Figure 5: D-level sentence distribution of human and robot language.

that were used more than once during dialogues is significantly ($p < 0.001$) higher in robot utterances (Mean = 15.8, SD = 31.4) compared to human utterances (Mean = 6.1, SD = 8.4). This means that the robot tends to use the same combinations of words repeatedly, while people do vary their language more. The majority of trigrams is also used just once by both people and the robot, although the proportion is quite different: 75% of human trigrams and only 65% of robot trigrams are used once in the dialogues. Those trigrams that are used more than once, have an average frequency of 15.8 (SD = 30.9) for robot, and only 4.4 (SD = 4.6) for human utterances.

The results of bigrams and trigrams analysis support the conclusion that human language in human-robot conversations is more rich, varied, and diverse than that of the robot. Figure 4 shows that poor lexical variation of a robot language is influenced a lot by the fact that the robot often uses the phrase "I am afraid I cannot help you with that", which may be said when the speech recognition confidence does not reach an adequate threshold, or when no known keywords are detected in human utterances. As Figure 4 shows, 7 out of 10 most frequent trigrams in a robot language are variations of that specific phrase.

### 4.3 Syntactic Variation and Discourse Phenomena

We used the D-Level Analyser (Lu, 2009) to evaluate syntactic variation and complexity of human references using the revised D-Level Scale (Lu, 2014). The scale has eight levels of syntactic complexity, where levels 0 and 1 include simple or incomplete sentences and higher levels include sentences with more complex structures.

Figure 5 shows a similar syntactic variation in human and robot language, although there are slight differences, e.g. people tend to use a higher percentage of both the simplest and the most complicated sentences. In general, the majority of all the sentences, used both by humans and by a robot, are simple sentences. This is because the topic of a human-robot conversation is quite simple and does not require a lot of complicated syntactic structures.

The results of initial linguistic analysis, together with results of analysis of multimodal signals, suggest that linguistic, as well as other multimodal features, may be important in predicting human perception of a robot. However, average scores can be misleading, as they only provide a system-level overview but do not measure the strength of association with human ratings. This led us to inspect the correlation between the ratings of the robot and all the multimodal features of a dialogue.

## 5 Correlation between Robot Ratings and Multimodal Features of Human-Robot Dialogue

A summary of correlation results is presented in Table 4. The results reveal that different groups of features correlate with different groups of human ratings. For example, emotional features, such as intensity of happiness, correlate strongly with perceived anthropomorphism of a robot, so that a person who more strongly expresses happiness during a dialogue with the robot probably perceives it as friendlier and nicer. Human ratings of perceived robot anthropomorphism also correlate with a lexical diversity of human language: people tend to use more diverse language when speaking to a robot that they perceive as conscious, natural, and humanlike (see an example in Table 5).

Average F0 value of human speech correlates strongly with perceived intelligence of the robot, specifically with a robot being more knowledgeable. The ratio of tasks per turn is, unsurprisingly, strongly correlated with perceived robot intelligence. The more dialogue turns people need to complete the same number of tasks, the more they perceive the robot as ignorant and unintelligent. Features of robot language also correlate with how it is perceived: the more words (including unique ones) per sentence it generates, the more competent it appears to humans.

It is interesting to notice that some linguistic

| Group of features | Features | Significant correlation with human ratings (Spearman) | |
| --- | --- | --- | --- |
| | | *Feature calculated for* Human | *Feature calculated for* Robot |
| Emotional features | Happiness | Friendly (0.72), Nice (0.67) | – |
| | Surprise | NA | – |
| | Sadness | Sensible (0.71) | – |
| Prosodic features | F0, avg | Knowledgeable (0.57) | – |
| | F0, max | NA | – |
| | F0, diff | NA | – |
| Non-linguistic dialogue-related features | Speech duration, sec | NA | NA |
| | No of turns | Intelligent (0.64), Knowledgeable (0.54) | Disliked (0.54) |
| | No of self-repetitions | Clear (0.65), Easy (0.52), Understandable (0.50) | Awful (0.58), Does not meet expectations (0.52), Annoying (0.49) |
| | No of completed tasks | Clear (0.63), Meets expectations (0.58) | – |
| | Tasks / Turn | Ignorant (0.58), Unintelligent (0.52) | Humanlike (0.63), Unintelligent (0.55), Ignorant (0.53) |
| Linguistic dialogue-related features | Utterance length | Responsible (0.54) | Annoying (0.62), Obstructive (0.50) |
| | Words / Utterance | NA | Annoying (0.62), Obstructive (0.57) |
| | Unique words / Utterance | NA | Annoying (0.62), Obstructive (0.57) |
| | Lexical Diversity | Conscious (0.61), Humanlike (0.58), Natural (0.56) | Supportive (0.54) |
| | No of sentences | Responsible (0.50) | Easy (0.50), Ignorant (0.57) |
| | Words / Sentence | NA | Competent (0.52) |
| | Unique words / Sentence | Confusing (0.54) | Competent (0.56) |

Table 4: Summary of correlation between a mean value of feature, calculated during a dialogue, and human ratings of the robot. Only significant correlations are included. "NA" means no significant correlation was observed, "–" means that correlation was not calculated.

H:  Where is Costa? Is it far? [LD = 0.83]
R:  A place we call "reality". [disfluency]
H:  Sorry I didn't understand. [LD = 1.0]
R:  There is no need to apologize. Perhaps I can explain it better. Which part are you confused about?

H:  Bye bye [LD = 0.50]
R:  My birthday is September 9th. [disfluency]
H:  Bye bye [LD = 0.50]
R:  Would you like to do a search for it? [disfluency]
H:  No. [LD = 1.0]

Table 5: An example of a dialogues that resulted in: Left - high human ratings for anthropomorphism, Right - low human ratings for anthropomorphism. Both dialogues present situations when human intent was not recognised by the robot. LD denotes lexical diversity.

and non-linguistic dialogue features correlate with different human ratings depending on whether the features are calculated for human or robot language. For example, a higher number of human turns during a dialogue correlates strongly with a robot being perceived as intelligent and knowledgeable, while a higher number of robot turns correlates with it being disliked. Longer human sentences show that a robot is perceived as more responsive, while longer robot sentences correlate with a robot being annoying and obstructive.

The results show that some features, observable during a human-robot dialogue, correlate strongly and significantly with different groups of human ratings. However, it is not obvious if a strong correlation also means that there is a causal relation-ship between human language or multimodal behavioural features and ratings of the robot. This leads us to inspect whether the previously discussed features may be used for predicting potential ratings.

## 6 Predicting Perception of Robots in Human-Robot Dialogue

In order to develop a model that predicts potential human ratings on robot likeability and perceived intelligence, we use the previously discussed prosodic features, dialogue-related characteristics, and detected emotional intensities, as predictive features of the model. For the prediction itself, we use ensemble learning (Random Forest, RF) (Breiman, 2001) which is a state-of-

| Group of rating | Emotions only | Prosody only | Non-linguistic only | Linguistic only | All combined | Baseline | Average rating |
|---|---|---|---|---|---|---|---|
| Likeability | **0.85** | 1.03 | 0.96 | 0.87 | 1.00 | 1.41 | 4.01 |
| Perceived Intelligence | 0.73 | 0.94 | **0.69** | 0.83 | 0.89 | 0.83 | 3.44 |
| Interpretability | 0.71 | 0.87 | **0.56** | **0.68** | 0.81 | 0.92 | 3.56 |

Table 6: Performance of prediction, calculated using root-mean-square error (RMSE). The results are averaged over all the ratings that belong to the group and outperform the baseline. Bold denotes the smallest average error and means the best predicted result of the model.

the-art algorithm that can be applied in a dynamic dialogue situation and is able to combine the respective strengths of different informative features into a single model.

**Setup:** We use a 70/30% split for training and testing and 10-fold cross-validation on the training data to tune the optimal number of predictors selected for growing trees. 100 trees were grown with 2 variables randomly sampled as candidates at each split. We investigate five different models used as predictors: 1) emotional intensities, 2) prosodic features, 3) non-linguistic dialogue features, 4) linguistic dialogue features, and 5) all the features combined.

**Results:** The results in Table 6 show that different groups of features are better predictors of different groups of ratings. For example, combining dialogue-related features only (either linguistic or non-linguistic) as predictors, produces the lowest root-mean-square error (RMSE) for many ratings out of the perceived intelligence and intelligibility groups. This means that a combination of dialogue-related features is producing the best prediction of such aspects of perceived robot intelligence as e.g. responsiveness, intelligence, or predictability.

Emotional features are shown to be the best predictors of some aspects of robot likeability and perceived anthropomorphism. For example, the ratings for *unconscious-conscious, unfriendly-friendly* or *awful-nice* are best predicted using emotional features only. Other aspects of robot anthropomorphism and likeability, such as *machinelike-humanlike* or *disliked-liked*, are best predicted by using only prosodic features of human speech as predictors. Combining emotional, prosodic and dialogue-related features rarely improves the results of rating predictions. In some cases, e.g. predicting the ratings for *unresponsive-responsive*, a combination of all the features pro-

duces the same results as dialogue-related features alone. In one case a combination of all features does improve prediction results, this is the rating showing if the robot meets human expectations or not. This is probably because human expectations consist of different aspects themselves: some expect the robot to be anthropomorphic and likeable, other prefer it to be intelligent and easily interpretable.

## 7  Discussion and Conclusions

In this paper, we show how dialogue features correlate with the user's perception of a robot (e.g. strong correlation between higher number of human turns and higher robot's perceived intellect, or between higher number of sentences per robot utterance and robot's perceived ignorance), as well as correlations between emotional features and robot likeability.

Using the findings described in this paper, a predictive model could be implemented using emotional intensities (*happiness, sadness, and surprise*) in order to better predict the user's perception of the robot. This model can provide valuable information on how to design more engaging dialogues between robots and humans. The combination of these emotional features, along with the dialogue-related features (both linguistic and non-linguistic) and the F0 value can also provide better feedback in cases where, for instance, a smile can create ambiguity of the perceived user's emotional display (Halpern and Kets, 2012).

In future work, these emotional features coming from real-time facial expression recognition could be used as an online estimator of how well or badly a dialogue is progressing, which would be an important component of a reward signal for Reinforcement Learning approaches to HRI.

## References

Christoph Bartneck, Dana Kulić, Elizabeth Croft, and Susana Zoghbi. 2009. Measurement instruments for the anthropomorphism, animacy, likeability, perceived intelligence, and perceived safety of robots. *International journal of social robotics* 1(1):71–81.

Leo Breiman. 2001. Random forests. *Machine learning* 45(1):5–32. https://doi.org/10.1023/A:1010933404324.

Felipe Cid, José Augusto Prado, Pablo Bustos, and Pedro Nunez. 2013. A real time and robust facial expression recognition and imitation approach for affective human-robot interaction using gabor filtering. In *Intelligent Robots and Systems (IROS), 2013 IEEE/RSJ International Conference on*. IEEE, pages 2188–2193.

Laurence Devillers, Marie Tahon, Mohamed A Sehili, and Agnes Delaborde. 2015. Inference of human beings emotional states from speech in human–robot interactions. *International Journal of Social Robotics* 7(4):451–463.

Paul Ekman. 2004. Emotional and conversational nonverbal signals. In *Language, knowledge, and representation*, Springer, pages 39–50.

C. Gardent, A. Shimorina, S. Narayan, and L. Perez-Beltrachini. 2017. Creating training corpora for micro-planners. In *Proceedings of the 55th Annual Meeting of the Association for Computational Linguistics (Volume 1: Long Papers)*. To appear.

Joseph Y. Halpern and Willemien Kets. 2012. Ambiguous language and differences in beliefs. *CoRR* abs/1203.0699. http://arxiv.org/abs/1203.0699.

Ota Herm, Alexander Schmitt, and Jackson Liscombe. 2008. When calls go wrong: How to detect problematic calls based on log-files and emotions? In *Ninth Annual Conference of the International Speech Communication Association*.

Bettina Laugwitz, Theo Held, and Martin Schrepp. 2008. Construction and evaluation of a user experience questionnaire. In *Symposium of the Austrian HCI and Usability Engineering Group*. Springer, pages 63–76.

Xiaofei Lu. 2009. Automatic measurement of syntactic complexity in child language acquisition. *International Journal of Corpus Linguistics* 14(1):3–28.

Xiaofei Lu. 2012. The relationship of lexical richness to the quality of esl learners oral narratives. *The Modern Language Journal* 96(2):190–208.

Xiaofei Lu. 2014. *Computational methods for corpus annotation and analysis*. Springer.

Raveesh Meena, José Lopes Gabriel Skantze, and Joakim Gustafson. 2015. Automatic detection of miscommunication in spoken dialogue systems. In *16th Annual Meeting of the Special Interest Group on Discourse and Dialogue*. page 354.

Jekaterina Novikova, Leon Watts, and Tetsunari Inamura. 2015. Emotionally expressive robot behavior improves human-robot collaboration. In *Robot and Human Interactive Communication (RO-MAN), 2015 24th IEEE International Symposium on*. IEEE, pages 7–12.

Ioannis Papaioannou, Christian Dondrup, Jekaterina Novikova, and Oliver Lemon. 2017. Hybrid chat and task dialogue for more engaging hri using reinforcement learning. In *Robot and Human Interactive Communication (RO-MAN), 2017 26th IEEE International Symposium on*. IEEE.

Ioannis Papaioannou and Oliver Lemon. 2017. Combining Chat and Task-Based Multimodal Dialogue for More Engaging HRI: A Scalable Method Using Reinforcement Learning. In *Proceedings of the Companion of the 2017 ACM/IEEE International Conference on Human-Robot Interaction*. ACM, New York, NY, USA, HRI '17, pages 365–366. https://doi.org/10.1145/3029798.3034820.

Javier G Rázuri, David Sundgren, Rahim Rahmani, Antonio Moran, Isis Bonet, and Aron Larsson. 2015. Speech emotion recognition in emotional feedback for human-robot interaction. *International Journal of Advanced Research in Artificial Intelligence (IJARAI)* 4(2):20–27.

Alexander Schmitt, Benjamin Schatz, and Wolfgang Minker. 2011. Modeling and predicting quality in spoken human-computer interaction. In *Proceedings of the SIGDIAL 2011 Conference*. Association for Computational Linguistics, pages 173–184.

Alexander Schmitt and Stefan Ultes. 2015. Interaction quality: assessing the quality of ongoing spoken dialog interaction by expertsand how it relates to user satisfaction. *Speech Communication* 74:12–36.

Adriana Tapus and Maja J Mataric. 2007. Emulating empathy in socially assistive robotics. In *AAAI Spring Symposium: Multidisciplinary Collaboration for Socially Assistive Robotics*. pages 93–96.

Myrthe Tielman, Mark Neerincx, John-Jules Meyer, and Rosemarijn Looije. 2014. Adaptive emotional expression in robot-child interaction. In *Proceedings of the 2014 ACM/IEEE international conference on Human-robot interaction*. ACM, pages 407–414.

Zhihong Zeng, Maja Pantic, Glenn I Roisman, and Thomas S Huang. 2009. A survey of affect recognition methods: Audio, visual, and spontaneous expressions. *IEEE transactions on pattern analysis and machine intelligence* 31(1):39–58.

# Towards Problem Solving Agents that Communicate and Learn

Anjali Narayan-Chen[1], Colin Graber[1], Mayukh Das[2], Md Rakibul Islam[3],
Soham Dan[1], Sriraam Natarajan[2], Janardhan Rao Doppa[3],
Julia Hockenmaier[1,5], Martha Palmer[4] and Dan Roth[1]
[1]University of Illinois at Urbana-Champaign, [2]Indiana University, [3]Washington State University,
[4]University of Colorado at Boulder, [5]Allen Institute for Artificial Intelligence

## Abstract

Agents that communicate back and forth with humans to help them execute non-linguistic tasks are a long sought goal of AI. These agents need to translate between utterances and actionable meaning representations that can be interpreted by task-specific problem solvers in a context-dependent manner. They should also be able to learn such actionable interpretations for new predicates on the fly. We define an agent architecture for this scenario and present a series of experiments in the Blocks World domain that illustrate how our architecture supports language learning and problem solving in this domain.

## 1 Introduction

An agent that can engage in natural, back-and-forth communication with humans to help them complete a real world task requires the ability to understand and produce language in the context of that task (i.e. to map between utterances and meaning representations the problem solving components of the agent can act on in a particular situation). The agent may also need to initiate clarification requests when communication fails, and to learn new domain (or conversation) specific vocabulary and its meaning. This kind of symmetric, grounded communication with a problem-solving agent goes significantly beyond the one-step, single direction understanding tasks considered in standard semantic parsing (e.g. Zelle and Mooney, 1996; Zettlemoyer and Collins, 2005; Clarke et al., 2010) or even short, simple instructions to robots (e.g. Tellex et al., 2011).

In order to focus on these concept learning and communication issues, we deliberately limit ourselves here to a simple, simulated environment.

We operate in a two-dimensional Blocks World domain where a human wants one or more shapes to be constructed on a grid. The human needs to communicate the goal of this planning task to the agent. Once the agent has understood the instructions and its planning is done, it communicates its plan to (possibly) another human who will then execute this plan. Depending on the complexity of the task and the linguistic capabilities of the agent, this scenario may require a lot of back-and-forth communication. If the human omits details from their description and prevents the agent from accomplishing the task, we expect the agent to initiate communication and ask clarification questions. If the human uses vocabulary that is new to the agent, we expect the agent to ask for a definition and the human to teach the agent its meaning in the domain.

We define an agent architecture named COG that allows us to investigate the challenges arising in this symmetric communication scenario. COG combines a problem solving (planning) component with a basic language understanding and generation system that is initially only equipped with a limited vocabulary. We perform a sequence of experiments of increasing complexity that illustrate how our architecture supports the problem solving scenarios described above, and how language learning is accomplished within this architecture. We argue that, within this architecture, all the agent's capabilities – comprehension, production, problem solving and learning – can be improved with additional communication.

Section 2 defines the domain and problem setup. Section 3 provides an overview of the COG architecture. Section 4 describes COG's current components (language comprehension, production, memory, problem solving, and dialogue mediation). Section 5 describes our experiments and results.

95

*Proceedings of the First Workshop on Language Grounding for Robotics*, pages 95–103,
Vancouver, Canada, July 30 - August 4, 2017. ©2017 Association for Computational Linguistics

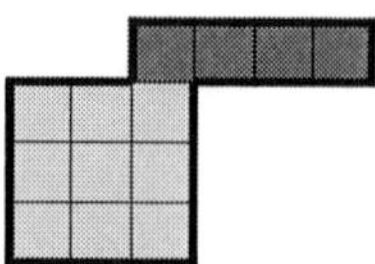

Figure 1: A complex shape, which can be viewed as conjunction of simpler known shapes: a row (dark, `width` = 4) and a square (light, `size` = 3).

## 2 Domain and Problem Setup

We consider a two-dimensional (2D) Blocksworld domain. There is a 2D grid and the goal is to build different configurations with square blocks based on their natural language descriptions. We assume agents come equipped with the vocabulary and definitions of primitive concepts, e.g. for simple shapes (*row*, *square*, etc.) and spatial relations (*adjacent*, *on top of*, etc.), but may have to learn the definitions of new terms and concepts that arise in communication with the human user.

We define three different types of goal descriptions and design corresponding tasks for evaluation. During evaluation, the agent must automatically identify the task and respond appropriately.

**Task 1: Complete descriptions.** For the first task, the agent is provided with a complete description, in natural language, of a target configuration consisting of one or two shapes. The definition is complete and does not require further clarification. This tests the ability of the agent to understand and ground specific descriptions.

**Task 2: Descriptions with missing information.** For the second task, the agent is again provided with a description of a target configuration consisting of one or two shapes; however, the description will not be specific enough for the agent to be able to ground it to a unique configuration. This tests the ability of the agent to recognize when some information is missing from a description and to initiate a dialogue which will clarify these details.

**Task 3: Descriptions of new shapes.** For the third task, the agent is asked to construct a complex shape which is not contained within its initial vocabulary of primitive concepts (e.g., the letter "L"). This tests the ability of the agent to extend its vocabulary through interaction with a human.

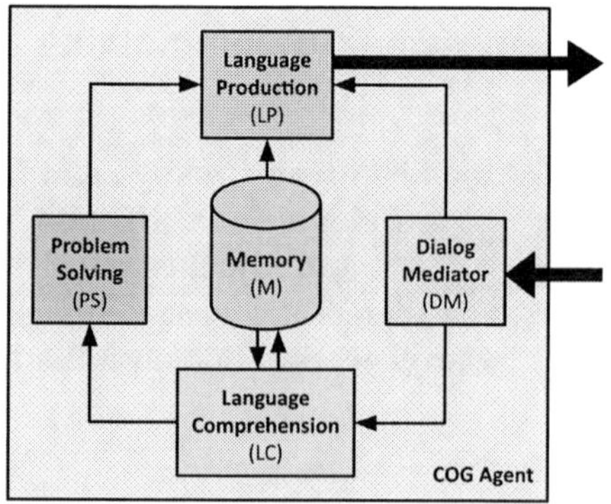



Figure 2: The agent architecture. Interaction (thick arrows) is done via the Dialogue Mediator (input) and the Language Production component (output). Thin arrows indicate the flow of information between the internal components.

## 3 The COG Agent Architecture

Agents that solve a task require explicit knowledge of the corresponding real world concepts. Agents that also communicate about solving tasks additionally need to be able to map linguistic expressions such as "square" or "on top of" to these concepts. They also need to know to which situations these concepts apply (e.g. whether a given configuration can be referred to as a square, or whether a row can be placed on top of that square). The representations used by the different components of our agent therefore vary in their specificity. This requires additional work to bridge the resulting representational gaps. To represent the definition of concepts, we store "lifted" representations, i.e. rules whose predicates only contain uninstantiated variables (e.g. rectangles have height and width). By contrast, problem solving requires fully grounded representations in which shapes are placed at specific locations on the grid (e.g., a 3 by 4 rectangle located at $(0,0)$). Language comprehension and production operate on a middle ground where some parameters may be instantiated, even if shapes are not necessarily placed at specific locations (e.g., a 3 by 4 rectangle).

Figure 2 describes the architecture of our agent. The **Language Comprehension (LC)** module converts natural language into an executable, grounded, declarative representation in a STRIPS-like language (Fikes and Nilsson, 1971) that the **Problem Solving (PS)** module can act on. PS returns partial or complete problem solving plans, also in this language. If grounding fails, LC produces queries that are sent to the **Language Production (LP)** module.

The LP takes these queries or the plans produced by PS and returns natural language output. The **Memory (M)** module stores lifted representations of built-in and acquired predicates that are used by LC and LP. The **Dialogue Mediator (DM)** is aware of the current dialogue state. It can augment the natural language input before transferring it to LC (e.g. if the human utters a number, DM might provide the context, letting LC know that the number refers to the size of a specific shape). DM is also used by LP to generate utterances that are appropriate for the current dialogue state. Instead of narrating the fully grounded, block-wise sequential plans produced by PS, LP may identify subsequences of steps that correspond to instantiations of more abstract concepts (e.g., a sequence of blocks placed horizontally is a "row"), and generates language at this more natural level.

## 4 Current Components of COG

In this section, we describe the current implementations of the different modules (language comprehension, memory, problem solving, language production, and dialogue mediation) in COG, noting that the architecture is flexible and allows for us to plug-in other implementations as needed.

### 4.1 Language Comprehension

The LC module consists of semantic parsing and language grounding components.

### 4.1.1 Semantic Parsing

Our semantic parser is implemented as a neural sequence-to-sequence model with attention (Bahdanau et al., 2014; Dong and Lapata, 2016; Jia and Liang, 2016). The model consists of two LSTMs. The first LSTM (the encoder) processes the input sentence $\mathbf{x} = (x_1, \ldots, x_m)$ token-by-token, producing a sequence of hidden states $\mathbf{h}^s = (h_1^s, \ldots, h_m^s)$ as output. The second LSTM (the decoder) models a distribution $P(y_i|y_1, \ldots, y_{i-1}; h^s)$ at each time step over output tokens as a function of the encoder hidden states and the previous outputs. The final parse $\mathbf{y} = (y_1, \ldots, y_n)$ is obtained by selecting the token at each time step that maximizes this probability and feeding a learned embedding for it into the LSTM as part of the next input. Multiple sentence inputs are processed sentence-by-sentence, where the initial encoder hidden state for a given sentence is set to the final encoder hidden state for the previous sentence, and the initial encoder hidden state for the first sentence is the zero vector. The final logical form is the conjunction of the logical forms for each individual sentence.

The parser is trained with a small fixed-size vocabulary; however, to represent new shapes it needs to be able to output new predicates for shapes that it has not encountered during training. We accomplish this by using an attention-based copying mechanism (Jia and Liang, 2016; Gulcehre et al., 2016; Gu et al., 2016). At every time step, the decoder may either output a token from the training vocabulary or copy a word from the input sentence. Hence, when new shapes are encountered in the input, the parser is able to copy the shape name from the input sentence to define a new predicate.

### 4.1.2 Language Grounding

Grounding is beyond the capabilities of a semantic parser, which may interpret a sentence such as "Place a square of size 4 on top of the rectangle" without any *understanding* of the key predicates (what is a "square"?) or relations (what is "on-top-of"?). LC therefore includes a grounding component which converts the output of the semantic parser into executable grounded representations that can be directly used by PS. This component obtains definitions of predicates from M and uses a geometric reasoner to identify feasible shape placements. The reasoner assigns location coordinates to each shape relative to the best bounding box it can find for the entire configuration, given the grid boundaries. A full description and its grounding process are given in Figure 3.

Grounding succeeds immediately if the agent is given a complete goal description, as in Task 1. Grounding fails if the goal description is incomplete or unknown, as in Tasks 2 and 3. In these cases, a clarification query $Q$ is issued and passed to LP.

Formally, a **complete goal description** of a target configuration is defined as a tuple $\mathcal{G} = \langle \{\langle s_i, id_i, \wedge_k d_i^{(k)} \rangle\}_{i \in S}, \wedge_{j \in [S \times S]} f_j \rangle$, where $S$ is the list of shapes and $s_i$, $id_i$ and $\wedge_k d_i^{(k)}$ are the shape name, identifier, and dimension attributes of shape $i$ respectively. $f$ encodes pair-wise

| **Human:** | "Build a square of size 3.  Then, construct a row of 4 blocks above the square. The left end of the row should be above the upper right corner of the square." |
| --- | --- |
| **Parse:** | square(a) ∧ size(a, 3) ∧ row(b) ∧ width(b, 4) ∧ ... |
| **Ground:** | square(0, 2, 3), row(2, 5, 4) |
| **Plan:** | (putdown b1 0.0 0.0), (putdown b2 1.0 0.0), (putdown b3 2.0 0.0), ... |

Figure 3: A fully-specified goal description for the configuration in Figure 1 and its path through the architecture. The geometric reasoner inferred that the bounding box's lower left corner is at $(0, 2)$, which is also the lower left corner of the square.

spatial relations between shapes. A complete goal description for the configuration in Figure 1 is $\langle\{\langle \texttt{row}, \texttt{a}, \texttt{width} = 4\rangle, \langle \texttt{square}, \texttt{b}, \texttt{size} = 3\rangle\}$, $\texttt{spatialRelation}(\texttt{a}, \texttt{b}, \texttt{upperRightCorner})\rangle$.

An **incomplete goal description** $\mathcal{G}^I$ is a goal description where the values of one or more dimensional or spatial attributes are missing: $\mathcal{G}^I = \mathcal{G} - x$, where $x = x_d \cup x_f$ (with $x_d \subseteq \{d_i\}_{i \in S}$ and $x_f \subseteq \{f\}$) is the missing information. In this case, a query $Q$ asks for values of the missing dimensions or spatial relations (i.e., $Q = x$). An incomplete goal description for Figure 1 is $\langle\{\langle \texttt{row}, \texttt{a}, \texttt{width} = \textbf{null}\rangle, \langle \texttt{square}, \texttt{b}, \texttt{size} = 3\rangle\}, \textbf{null}\rangle$. Here, the width of the row and the spatial relation between the shapes are unknown.

An **unknown goal description** $\mathcal{G}^U$ occurs when one or more of the shapes in a goal description are not known to the agent (i.e., the memory module M does not contain their definitions): $\mathcal{G}^U = \mathcal{G}$ if $\exists s_i \colon s_i \in S, s_i \notin$ M. In this case, a query $Q$ asks for additional clarification about the unknown concepts (i.e., $Q = \textit{Define}(s_i)$). Figure 1 could be described with a new concept Foo $\notin$ M: $\mathcal{G} = \texttt{Foo}(\texttt{p}) \wedge \texttt{dim}_1(\texttt{p}, 3) \wedge \texttt{dim}_2(\texttt{p}, 4)$.

### 4.2  Memory

A key challenge for communicating agents is the necessity to learn the interpretation of new concepts (e.g. names of unknown shapes, spatial relations, or actions)[1] that arise during the communication with the human. In our agent, the Memory module stores lifted representations of these acquired concepts that are parameterized for the configuration's dimensions, and hence generalize beyond the specific instances encountered by

---

[1] We currently restrict ourselves to unknown shapes.

| **Human:** | "Build a 3 by 4 Foo." |
| --- | --- |
| **Parse:** | Foo(p) ∧ dim₁(p, 3) ∧ dim₂(p, 4) |
| **Query:** | *Define*(Foo) |
| **System:** | "Sorry, I don't know what 'Foo' is. Could you describe how to build the Foo for me using rows, columns, squares, and rectangles?" |
| **Human:** | "Build a square of size 3.  Then, construct a row of 4 blocks above the square. The left end of the row should be above the upper right corner of the square." |
| **Lift:** | Foo(p) ∧ dim₁(p, ?d₁) ∧ dim₂(p, ?d₂) → square(a) ∧ size(a, ?d₁) ∧ row(b) ∧ width(b, ?d₂) ∧ spatial−rel(... |

Figure 4: A dialogue triggered by the unknown word "Foo" for the configuration in Figure 1. LC issues a query that prompts a request to define "Foo" in terms of known shapes. This definition is first parsed and grounded as in Figure 3, then learned by being lifted and stored in M ($?d_k$ identifies dimension variables).

the agent. When the agent receives an unknown goal, it asks the human for a definition of the unknown concept and expects a new goal description $\mathcal{G}^N$ that defines it in terms of known concepts. This definition is then added to the agent's domain knowledge stored in M.

Learning a new concept is done via a "lifting" process, as follows:

1. When an unknown goal description $\mathcal{G}^U$ is received, LC issues a query $Q^U$ which is then realized and posed to the human by LP.

2. LC receives a natural language response containing a new goal description $\mathcal{G}^N$. If $\mathcal{G}^N$ contains any unknown concepts, the previous step must be called recursively. Once $\mathcal{G}^N$ is complete, it is grounded and passed to PS.

3. If a successful plan was generated for $\mathcal{G}^N$, the concept declaration and goal definition are lifted by converting the given dimension and relative location constants to variables while ensuring parameter sharing between the new concept and its definition. A mapping is created and inserted into M. Lifting ensures generalization over arbitrary examples of the new learned concept.

Figure 4 illustrates an example of a description that elicits a query from the system for further information and the subsequent resolution process. Our present implementation of lifting is restrictive. The challenges in handling the general learning setting and potential principled approaches will be discussed in later sections.

### 4.3 Problem Solving

The problem solving module reasons about the configurations communicated to it as conjunctive logical expressions and generates a set of problem solving steps (plan) to achieve the given target configuration as the output. The problem solving module proceeds in an anytime fashion and generates a partial plan as far as possible. We employ a Hierarchical Task Network (HTN) planner (Erol et al., 1994; Nau et al., 2003; Ghallab et al., 2004) that searches for plans in a space of task decompositions. HTN allows for reasoning over different levels of abstraction, tasks and sub-tasks as well as primitive actions and is more intuitive for humans.

### 4.4 Dialogue Mediation

The role of the dialogue mediator in our framework is to guide the interaction with the human and delegate the tasks of parsing, planning, and query realization to the LC, PS, and LP modules respectively. The DM interacts with a GUI framework that allows for back-and-forth textual interaction between the human and the system as well as a visualization component that displays the output of the problem solver as block configurations on a grid. Via this framework, the DM accepts user-described goal descriptions and prompts the user to reword their utterances, clarify missing information, and define new shapes as needed.

The DM is also responsible for keeping track of the cumulative information gained about a goal configuration over a dialogue sequence in order to backtrack the states of the semantic parsing and problem solving components if mistakes occur during the interaction. In the semantic parser, backtracking consists of restoring the hidden states of the parser that were seen at a particular time of the interaction before the mistake was made. Backtracking in the problem solver involves deleting or modifying items in its goal description $G$. But, since the parser conditions hidden states on all previously seen sentences, the DM is not able to selectively delete or replace information at arbitrary points in the timeline of the dialogue. Hence, our experiments process goal descriptions on a sentence-by-sentence basis, allowing for clarifying questions to be made and resolved per-sentence.

Our current agent uses a rule-based dialogue mediator (implemented as a finite state machine), which alternates between four sets of states:

1. **Goal description parsing and planning.** Given an input user description of a goal configuration, DM passes this input through the LC and PS pipeline, backtracking and requesting a rewording if either module encounters a failure.

2. **Querying for and resolving clarifications.** When LC returns a a query $Q$ asking for values of missing dimensional or spatial features, DM requests that information from the user via the LP module (e.g., "What is the width of the row?"). The user may respond with a well-formed sentence describing the missing feature value ("The width of the row is 4") or with a fragment containing the desired information ("It's 4," "4," "Width is 4"). Given the context of the original query $Q$, DM extracts the value via a heuristic and reforms the user input into a well formed sentence, then returns to State 1 to handle the updated description.

3. **New shape learning.** When LC returns a query $Q$ indicating an unknown shape or concept, DM requests the user to describe the desired configuration using known shapes (i.e., rows, columns, squares, and rectangles). The description handling process proceeds regularly as in State 1 until the user indicates they are finished defining their new concept. This triggers the learning process to lift and store the definition in M; the new concept can then immediately be used like any known concept in future user descriptions.

4. **Shape verification.** Every time the plan for an individual shape in a configuration has been resolved, DM outputs the plan to the visualization component and asks the user if the configuration up until that point in the dialogue is correct. If the user indicates that something is wrong, DM removes the entire shape from the history of the parsing and planning components and asks the user to retry describing the shape from scratch. Once the shape has been verified, however, no further modifications may be made to that shape.

| # Shapes | Task 1 | Task 2 |
|:---:|:---:|:---:|
| 1 | 100% | 100% |
| 2 | 64% | 52% |

Table 1: Accuracies for Task 1 and Task 2.

For our experiments, a simplified version of this dialogue mediator was used to feed examples through the system pipeline. More details can be found in Section 5.

## 4.5 Language Generation

Our current system uses a predominantly template-based approach to generate user queries. For clarification questions, we use a simple rule-based approach for clarifying questions that increases the level of question explicitness if the user continually fails to understand the question or respond in an appropriate manner. For example, if the query $Q$ requests the width of a particular row, clarification questions may range from the very contextual "What's the width?" to the fully explicit "What is the width of the row you are currently building?"

## 5 Methodology and Experiments

Although we ultimately wish to evaluate our agent's live interactions with real human users, our current experiments feed it synthetically generated descriptions. For the descriptions with missing information, we randomly choose dimensional and/or spatial information to omit for every shape, providing the missing information in a follow-up response upon system query. We also generate sentences with descriptions with unknown terms that have to be clarified in further interactions. After a full interaction has been carried out (or halted early due to system comprehension failure), we automatically verify the system's output configuration against the gold configuration.

Input consisting of sentences describing various block configurations and their corresponding logical forms was produced using a template-based generation system. A subset of this data was used for training the semantic parser; the rest was used for the following experiments. The generation system was set up to create configurations randomizing over shape types, sizes, and relative locations, and the templates for the sentences were designed to introduce lexical and grammatical variety. Overall, the generation system is able to pro-

| Human: | Build a square of size 3. Then, construct a row of 4 blocks above the square. The left end of the row should be above the upper right corner of the square. |
|---|---|
| System: | Is this okay so far? (y/n) |
| Human: | yes |
| System: | Great! |

Figure 5: Task 1 (Complete descriptions)

duce tens of thousands of different configuration descriptions.

Generated descriptions contain one sentence (for single shapes) and up to three sentences (for two shapes). The sentences have an average length of approximately ten words. Multi-sentence descriptions pose special challenges for semantic parsing due to the need to resolve coreference across sentences. Thus, increasing the number of shapes in a configuration dramatically increases the complexity of the parsing task.

The COG system was evaluated on three separate tasks as outlined in Section 2; we present the results below.

## 5.1 Task 1: Complete Descriptions

In the first part of the experiment, we test COG's ability to handle complete descriptions of configurations (see Figure 5 for an example dialogue). The parameter that we change in the course of this experiment is the number of primitive shapes used to build the configuration, which we vary from 1 to 2. For each parameter setting we have 50 test instances; this totals 100 test instances of complete descriptions with no missing information. COG's accuracy in interpreting these descriptions is shown in the left column of Table 1. The drop in accuracy in moving from one to two shapes is due to the drop in performance of the semantic parser when dealing with complex descriptions.

## 5.2 Task 2: Missing Information

In this experiment, we test COG's ability to process descriptions with missing information (see Figure 6 for an example dialogue). The system handles these situations by asking clarification questions, obtaining responses and building the configuration by incorporating these subsequent responses. Again, we vary the number of primitive shapes in the configuration from 1 to 2 and have 25 test instances for each case. Further, each primitive shape could potentially have either or both the dimension and relative spatial position missing in

| | |
|---|---|
| **Human:** | Form a column. |
| **System:** | What is its height? |
| **Human:** | 5 blocks |
| **System:** | Is this okay so far? (y/n) |
| **Human:** | yes |
| **System:** | Cool! Please continue to describe your configuration. |
| **Human:** | Finally, form a vertical line of size 5 to the right of the column. |
| **System:** | Can you describe the second column's location relative to the first? |
| **Human:** | Ensure that the top end of the vertical line is next to the third row of the column. |
| **System:** | Is this okay so far? (y/n) |

Figure 6: Task 2 (Missing information)

| | |
|---|---|
| **Human:** | Construct a balloon of height 3 and width 4. |
| **System:** | Sorry, I don't know what 'balloon' is. Can you describe how to build the balloon for me using rows, columns, squares, and rectangles? |
| **Human:** | Build a column of height 3. Then, build a 4 by 4 square above the column. Ensure that the second column of the square is above the top end of the column. |
| **System:** | Is this okay so far? (y/n) |
| **Human:** | yes |
| **System:** | Great! I've now learned what a balloon is. |
| | . . . |
| **Human:** | Build a balloon of height 2 and width 7. |
| **System:** | Is this okay so far? (y/n) |

Figure 7: Task 3 (Learning new shapes)

the initial description. For example, in a configuration involving 2 shapes, there could potentially be 3 pieces of information missing. The 25 test instances are generated such that the number of missing information pieces in a single instance is uniformly random in $[1, 3]$. COG's accuracy in interpreting incomplete descriptions is shown in the right column of Table 1. As before, the drop in accuracy in moving from one to two shapes is due to the drop in performance of the semantic parser when dealing with complex descriptions. The accuracy is further impacted by the interaction, there are more opportunities for the parser to misinterpret a natural language utterance.

### 5.3 Task 3: Learning New Shapes

In the final experiment, we evaluate COG's ability to learn new shapes (see Figure 7 for an example dialogue). We teach it five descriptions of new shapes, the results of which are presented in Table 2. Each new shape is defined using descriptions consisting of primitives (e.g., rows, columns, squares, and rectangles) of certain dimensions. We then tested the ability of the system to generalize and build other instances of the new shapes with altered dimensions. The "Outcome" column indicates whether the system was able to correctly lift all of the parameters of the input shape. Note that, when the dimension of a new shape consists of the sum of two dimensions (as in the third and fifth examples), parameter lifting fails for that dimension.

## 6 Discussion and Future Work

The goal of this paper was to present and study a new agent architecture that supports natural, symmetric communication with humans in the context of performing a real world task. Even in the simulated environment considered here, we had to address a number of challenges. First, language understanding to support task execution differs from e.g. standard semantic parsing in that it requires grounded meaning representations that can be executed and whose grounding may depend on the particular situation. Second, agents needs to be able to identify, request, and incorporate missing information that prevents task execution. Finally, agents need to be able to identify and learn the interpretation of new terms introduced during the interaction with the user.

Our current implementation has a number of obvious shortcomings. A primary bottleneck of COG is the semantic parser. If the input descriptions are not parsed correctly, there is very little the system can do to recover beyond asking for a rephrasing from the human. This creates problems when we attempt to experiment with complex configurations that involve three or more shapes. In these cases, the parser struggles to correctly identify the spatial relations for the third shape. Future work needs to address how to better handle longer and more complex descriptions.

Additionally, as the performance of Task 2 indicates, the dialogue mediator sometimes has problems translating partial responses into complete sentences that the parser can handle robustly. More work is required to develop a better treatment of implicit arguments, coreference, and discourse referents that are commonly present in these types of responses.

The generation component is also limited. For example, we currently cannot produce instructions for new shapes. We have also de-

| Initial Descriptions | Further Explanations | Outcomes |
| --- | --- | --- |
| Box of width 5 and height 3 | Construct a rectangle of width 5 blocks and height 3 blocks. | All parameters lifted correctly |
| Balloon of width 4 and height 3 | Build a column of height 3. Then, build a 4 by 4 square above the column. Ensure that the second column of the square is above the top end of the column. | All parameters lifted correctly |
| Balloon of width 4 and height 7 | Build a column of height 3. Then, build a 4 by 4 square above the column. Ensure that the second column of the square is above the top end of the column. | Failed to lift height of balloon |
| L of height 6 and width 3 | Build a column of 6 blocks. Then, build a row of 3 blocks to the right of the column. Make sure that the left end of the row is to the right of the bottom end of the column. | All parameters lifted correctly |
| L of height 6 and width 4 | Build a column of 6 blocks. Then, build a row of 3 blocks to the right of the column. Make sure that the left end of the row is to the right of the bottom end of the column. | Failed to lift width of L |

Table 2: Task 3 results for learning new shapes. We initially provide descriptions containing new shape terms with their parameters (left column). When prompted for further clarification, we then provide the descriptions in the middle column. After the shapes were learned, we instructed COG to build more instances of these new shapes while varying the size parameters to test how well the lifting worked; the result of this is in the right column.

veloped a grammar-based realizer inspired by OpenCCG (White and Baldridge, 2003; White, 2006) that operates over the first-order semantic representations used by our agent. We plan to augment the realizer's semantic lexicon with the learned definitions of predicates for new shapes, allowing our system to generate natural language instructions describing the new configurations.

One of the key challenges of the scenario we envision (and a fundamental problem in language acquisition) is the necessity to generalize across situations. This is required in order to learn general concepts from a few specific instances. At this point, our agent is able to generalize from a single example, but our learning mechanism is rather naive, and we can only handle simple parameter lifting from the primitive components of the new shape. To illustrate this, consider an example where we want to teach the system the shape "balloon". To do so, we must specify the dimensions of the square and column separately; consequently, the system will learn that the height of a "balloon" corresponds only to the height of the column (see Table 2). However, ideally we would like the system to learn a parameterization of "balloon" where the balloon's height corresponds to the sum of the square and column heights.

Our next step will focus on improving this mechanism. One possible direction involves the use of Inductive Logic Programming to induce high-level hypotheses from observations and background domain knowledge. One key challenge here, beyond learning, is a way to incorporate more complex learned hypotheses into our architecture in such a way that other components in our system can make use of it. A second challenge is that the agent is learning from a human that has limited knowledge, and little patience, so we cannot expect to see a large number of examples. We will consider the use of probabilistic logic models which can handle both issues by explicitly including the trade-off in the optimization function (Odom et al., 2015).

A final challenge is the application of our agent to new domains. Currently, the memory module contains all the knowledge required to plan and produce comprehensible responses. This declarative approach should generalize well to some simple enough domains, but will need to be extended to deal with more involved tasks and domains.

## Acknowledgements

This work was supported by Contract W911NF-15-1-0461 with the US Defense Advanced Research Projects Agency (DARPA) Communicating with Computers Program and the Army Research Office (ARO). Approved for Public Release, Distribution Unlimited. The views expressed are those of the authors and do not reflect the official policy or position of the Department of Defense or the U.S. Government.

# References

Dzmitry Bahdanau, Kyunghyun Cho, and Yoshua Bengio. 2014. Neural machine translation by jointly learning to align and translate. *arXiv preprint arXiv:1409.0473* .

J. Clarke, D. Goldwasser, M. Chang, and D. Roth. 2010. Driving semantic parsing from the world's response. In *Proc. of the Conference on Computational Natural Language Learning (CoNLL)*. http://cogcomp.cs.illinois.edu/papers/CGCR10.pdf.

Li Dong and Mirella Lapata. 2016. Language to logical form with neural attention. In *Proceedings of the 54th Annual Meeting of the Association for Computational Linguistics (Volume 1: Long Papers)*. Association for Computational Linguistics, Berlin, Germany, pages 33–43. http://www.aclweb.org/anthology/P16-1004.

Kutluhan Erol, James Hendler, and Dana S Nau. 1994. HTN planning: complexity and expressivity. In *Proceedings of the Twelfth AAAI National Conference on Artificial Intelligence*. pages 1123–1128.

R. E. Fikes and N. Nilsson. 1971. STRIPS: A new approach to the application of theorem proving to problem solving. *Artificial Intelligence* 2(3-4):189–208.

Malik Ghallab, Dana Nau, and Paolo Traverso. 2004. *Automated planning: theory & practice*. Elsevier.

Jiatao Gu, Zhengdong Lu, Hang Li, and Victor O.K. Li. 2016. Incorporating copying mechanism in sequence-to-sequence learning. In *Proceedings of the 54th Annual Meeting of the Association for Computational Linguistics (Volume 1: Long Papers)*. Association for Computational Linguistics, Berlin, Germany, pages 1631–1640. http://www.aclweb.org/anthology/P16-1154.

Caglar Gulcehre, Sungjin Ahn, Ramesh Nallapati, Bowen Zhou, and Yoshua Bengio. 2016. Pointing the unknown words. In *Proceedings of the 54th Annual Meeting of the Association for Computational Linguistics (Volume 1: Long Papers)*. Association for Computational Linguistics, Berlin, Germany, pages 140–149. http://www.aclweb.org/anthology/P16-1014.

Robin Jia and Percy Liang. 2016. Data recombination for neural semantic parsing. In *Proceedings of the 54th Annual Meeting of the Association for Computational Linguistics (Volume 1: Long Papers)*. Association for Computational Linguistics, Berlin, Germany, pages 12–22. http://www.aclweb.org/anthology/P16-1002.

Dana S Nau, Tsz-Chiu Au, Okhtay Ilghami, Ugur Kuter, J William Murdock, Dan Wu, and Fusun Yaman. 2003. SHOP2: An HTN planning system. *Journal of Artificial Intelligence Research* 20:379–404.

P. Odom, T. Khot, R. Porter, and S. Natarajan. 2015. Knowledge-based probabilistic logic learning. In *Proceedings of the Twenty-Ninth AAAI Conference on Artificial Intelligence*. pages 3564–3570.

Stefanie Tellex, Thomas Kollar, Steven Dickerson, Matthew R. Walter, Ashis Gopal Banerjee, Seth J. Teller, and Nicholas Roy. 2011. Understanding natural language commands for robotic navigation and mobile manipulation. In *Proceedings of the National Conference on Artificial Intelligence (AAAI)*.

Michael White. 2006. Efficient realization of coordinate structures in Combinatory Categorial Grammar. *Research on Language & Computation* 4(1):39–75.

Michael White and Jason Baldridge. 2003. Adapting chart realization to CCG. In *Proceedings of the 9th European Workshop on Natural Language Generation*. pages 119–126.

J. M. Zelle and R. J. Mooney. 1996. Learning to parse database queries using inductive logic proramming. In *Proceedings of the National Conference on Artificial Intelligence (AAAI)*. pages 1050–1055.

L. Zettlemoyer and M. Collins. 2005. Learning to map sentences to logical form: Structured classification with probabilistic categorial grammars. In *Proceedings of Uncertainty in Artificial Intelligence (UAI)*.

# Author Index

**Association for Computational Linguistics**
209 N. Eighth Street
Stroudsburg, Pennsylvania 18360